Motor Cycle Tuning
(Two-Stroke)

Motor Cycle Tuning (Two-Stroke)

Second Edition

John Robinson

OXFORD   AUCKLAND   BOSTON   JOHANNESBURG   MELBOURNE   NEW DELHI

An imprint of Butterworth-Heinemann
Linacre House, Jordan Hill, Oxford OX2 8DP
225 Wildwood Avenue, Woburn, MA 01801-2041
A division of Reed Educational and Professional Publishing Ltd

℞ A member of the Reed Elsevier plc group

First published 1986
Reprinted 1988, 1989, 1990, 1992
Second edition 1994
Reprinted 1995, 1996, 1997, 1999, 2000, 2001

ISBN 0 7506 1806 X

Printed and bound in Great Britain by MPG Books Ltd, Bodmin, Cornwall

PLANT A TREE

FOR EVERY TITLE THAT WE PUBLISH, BUTTERWORTH-HEINEMANN
WILL PAY FOR BTCV TO PLANT AND CARE FOR A TREE.

Contents

Preface

From Post Office delivery bike to world championship winner; from economy runabout to monsters which put out well over 300 bhp/litre. That's the story of the post-war two-stroke, and the only competition classes which aren't dominated by them are those whose formulae virtually exclude these simple engines, 'with only three moving parts'.

And still development continues. There are motocrossers with more violent engines than the road racers of a few years ago, and street bikes which have as much performance as their GP ancestors – more if you allow for their flexible engines and efficient silencing.

The great attraction of the two-stroke is its enormous potential, contrasted with its appealing simplicity. It is true that, armed with little more than a set of files, one can make profound changes to the power output of a two-stroke. Whether these changes will *increase* the power depends upon the understanding of the man who holds the file. After a while it becomes an intellectual challenge, but it doesn't alter the fact that a stock roadster engine can be turned into a machine capable of winning open-class races, for an outlay which is positively low by racing standards.

In this book I have attempted to put together my understandings of two-stroke development, both from direct experience and from a variety of more reliable sources.

J.R.

Acknowledgements

This book used up a lot of research, over a period of several years. During this time, many people were kind enough to help by providing information, data, illustrations and components for reference. In particular I'd like to thank the following: AE (Hepolite); Amal Carburetters; Martyn Barnwell; Dell'Orto Carburettors; Arnold Fletcher (Len Manchester Motorcycles); Patrick Gosling; Bernard Hargreaves (Harpowa); Heron Suzuki GB; Honda R & D; Honda UK; Kawasaki Motors UK; Mikuni Carburettors; Motor Industry Research Association; Mitsui Machinery Sales; Leon Moss (LEDAR); Lotus Engineering; NGK; Performance Bikes; Yamaha Motor NV; Yamaha R & D; Rod Sloane; and, of course, the development engineers at DKW.

J.R.

Chapter 1
Basic principles

Piston engines have two main functions. First they must work as pumps to draw in air and pass it through their working systems. Second, they must be able to extract heat from this gas flow and convert it into a usable form of energy.

Tuning means altering one or both of these processes in order to get more power output in certain conditions. It can mean the same as tuning a musical instrument, in the sense that a component, like a carburettor, can be adjusted and corrected until it is exactly right. It can also mean working on a part to improve its efficiency – or (more usually) to improve its efficiency within a certain speed range. Finally, tuning can mean making fairly coarse changes – like a bigger engine will produce more power than a smaller one.

It is not difficult to take a standard, production two-stroke and persuade it to make more power but it is essential to recognise that this is highly unlikely to make it a better engine, unless you have reason to suppose that your facilities and knowledge are better than the factory's. Engines, particularly roadster units, are designed with many compromises. As well as giving good power they have to be flexible and capable of being used by unskilled riders. They have to be cheap to make and still be reliable. They have to have a long service interval in a wide variety of conditions. Finally they have to meet a variety of emission controls and other legal limitations.

Because of these many and often conflicting demands, it is possible to increase peak power at the expense of some less important factors. A racer can make more noise than a roadster, it needs less flexibility, running costs are less critical and it can be overhauled at very frequent intervals. A full race engine will probably need one or two major overhauls per season (down as far as reconditioning the crankshaft) and will need top-end overhauls every three or four meetings.

One school of thought says that a racing engine should blow up as it crosses the finish line, to guarantee that it has been stressed to the limit. Horse-power is expensive.

There is one area in which the individual has an advantage over the production line. He can afford to spend time and effort on the engine, making sure that all the working clearances are just right and that parts line up exactly so that the gas passageways are smooth and are not interrupted by casting flashes or by protruding gaskets. Unfortunately, where Japanese

1

engines are concerned, this isn't much of an advantage. They are very good at getting the important details right and where the assembly might look a bit slapdash, you'll probably find that it doesn't matter.

Having said all that, two-strokes are still open to fairly extensive changes. It is possible to raise the output of a roadster by 20 to 30 per cent, if you don't mind having a motor with 'peaky' power characteristics. It can be raised by 50 per cent if you can accept a serious drop in reliability and the use of expensive racing components.

The expressions *power, torque, power-band* are going to crop up frequently, so it would be a good idea to get them defined straight away.

The output of an engine is felt, ultimately, as a force along the line of the drive chain. It is engine *torque* which is responsible for this force; torque is a force applied about a pivot (or, where the engine is concerned, a shaft). The further the force is moved from the pivot, the greater the torque – which is why a light person sitting on the end of a see-saw can balance a heavy person sitting part-way along the other side.

Engines develop torque, but in a shaft which is continuously moving, and this brings in a new concept: speed. If the engine were connected to a winch, a certain level of torque would be required to lift a particular weight. An engine which gave the same torque but at a higher speed would be able to lift the same weight but would do it faster. This gives rise to the notion of *power* – it is defined as torque multiplied by speed. One horsepower is 33,000 lb-ft per minute; one watt is 1 newton-metre per second.

If an engine gave the same torque at all speeds, the power would rise progressively as the speed went up (Fig. 1, section C). If the torque also rises as the speed increases then the power would rise more and more quickly (A

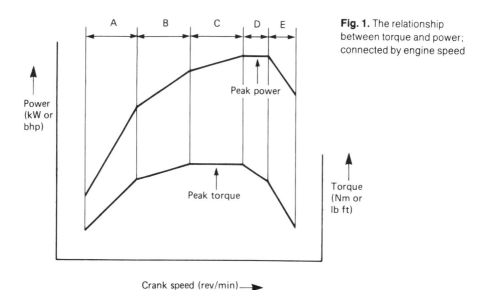

Fig. 1. The relationship between torque and power; connected by engine speed

and B). If the torque falls at a gentle enough rate, the power will remain constant (D) because the loss in torque will be balanced by the gain in speed. When the torque falls at a greater rate then the power will also fall (E).

When the speed of an engine is increased progressively from lowest to highest (with the throttle wide open), the torque usually starts fairly low, builds up gradually to a peak and then falls away, the decrease becoming more rapid as the engine approaches its maximum speed. At this upper speed limit the engine is moving so quickly that the ports are not open long enough to allow in enough gas to make as much torque as at lower speeds. Also there isn't as much time in which to burn it. The engine becomes inefficient both as a pump and as a heat exchanger.

The point at which the torque reaches a peak is where the engine is most efficient as a pump. In this speed region it will draw in more air during each cycle than at any other speed – and this concept of air flow is important to the tuner.

At lower speeds the pumping efficiency is not so good because its action depends to some extent on making the gas reach a suitable velocity so that its momentum can be used to help fill the engine. There is also a fair amount of overlap between the ports and if the engine is operated too slowly there will be time for gas which would normally be trapped inside to be pumped out through ports which are still open. The sizes of the ports and the time for which they are kept open have to be matched to the speed at which the engine has to run.

This gradual rise, peak and descent of the torque curve produces a power curve which, at low to medium engine speed, will rise quite sharply, continuing on to its own peak and fall. Sometimes two-strokes have torque curves which drop so sharply after peaking that peak power happens at the same speed as peak torque. If the high-speed torque doesn't drop too sharply, then peak power will happen at a higher engine speed.

The principles of tuning are: to increase the level of torque, which gives a proportionate increase in power without altering engine speed; or to maintain the same torque value but at a higher engine speed, which will also give more power and move the operating region further up the speed scale. A more extreme alternative is to use less torque but to do it at vastly increased engine speed, so that the increase in speed more than compensates for the reduction in torque and the overall change is a gain in horsepower.

Note that any alterations which make the ports etc., a better match for high engine speeds will make them more of a mismatch at low engine speeds. As the manufacturer has usually found the best combination of peak power and speed range, the result of an increase in peak power is usually that the speed range is reduced. Apart from reliability problems associated with increased speeds, there can be practical difficulties with an engine which only produces useful power over a narrow speed range. The gear ratios may be so widely spaced that even when the engine is taken to maximum speed in one gear, shifting up to the next will cause the revs to drop below the bottom limit of the power band. The program 'RL' in the Appendix shows how the power band and gear ratios can be matched. Also, if the engine is very inefficient at low speeds, starting and idling may become difficult.

A detailed look at the two-stroke cycle – and some of its variations – will show how an engine can be tuned and how different power characteristics can be produced.

Gas flow

The rising piston is used to draw air and fuel into the crankcase. The intake port can be opened and closed by the piston skirt (piston-ported) or by a rotary valve (disc valve) which is turned with the crankshaft. The program 'Piston' in the Appendix demonstrates this graphically. An alternative method is to have the flow of intake gas controlled by a reed valve, which opens when the crankcase pressure is low, and springs closed when the pressure rises to atmospheric.

As the piston descends (driven by the power stroke from the preceding cycle), it compresses the new gas in the crankcase; first the intake port has to be closed, to prevent blow-back through the carburettor and then the exhaust port in the cylinder has to open. This port is controlled by the top of the piston and this phase (exhaust open, gas being compressed in crankcase) is called exhaust blowdown. The hot gas at high pressure above the piston pops out violently into the exhaust port, and the pressure in the cylinder falls. At the same time crankcase pressure is increasing and as it becomes greater than cylinder pressure, the piston opens more ports, called the scavenge or transfer ports. These connect the crankcase below the piston with the cylinder above it, and the fresh gas in the crankcase is now pumped up through them, into the cylinder, displacing the remaining burnt gas into the exhaust.

The piston rises and after closing the scavenge and the exhaust ports it compresses the fuel/air mixture while also drawing in more fuel/air mixture

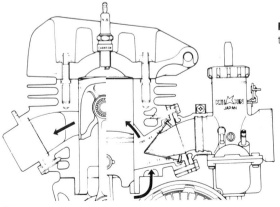

Fig. 2. Cylinder layout of a typical two-stroke and the gas flow through it

into the crankcase, ready for the next cycle. The compressed gas is ignited and the release of heat raises its pressure, forcing the piston down.

There have been several variations on this theme. The 1930s DKWs used various intake and scavenge mechanisms, including the pump-loading engine in which one cylinder was used to induct new gas and force it into a neighbouring cylinder for combustion. Several blower-scavenged engines have been used in which the air is supplied directly from a separately-driven supercharger. Another theme is the stepped piston design, in which the lower half of the piston has a greater bore than the top half and this is used to scavenge the cylinder, instead of the crankcase displacement.

The *volumetric efficiency* of an engine is the ratio of the volume of gas drawn into the cylinder divided by the volume swept by the piston. Because a two-stroke pumps from its crankcase into the cylinder, the amount of gas, compared to that displaced by the piston, is called the *delivery ratio.* This gas can also be lost into the exhaust, so what is left in the cylinder, divided by the swept volume gives the engine's *trapping efficiency.*

Intake
The intake phase needs to begin when crankcase pressure is lower than atmospheric – which determines the opening point of a disc valve engine. The port needs to close when the crankcase pressure goes above atmospheric – and this depends much more on engine speed. The weakness of piston-ported engines is that the opening and closing points have to be symmetrical about TDC. Disc valve engines can open and close wherever the designer chooses, and they are easy to modify.

Reed valve engines (in which a valve with flexible petals prevents blowback through the intake) give, in effect, variable timing through the entire speed range.

Originally developed to prevent blowback, the reeds allowed timing to be selected for high speed use, without suffering a major loss at low speeds. This worked well on kart engines and on trials bikes where low-speed flexibility and a wide power band were important. It didn't do much at high speed because of the physical restriction of the reed block. Developments mainly by Yamaha changed this, until eventually the reed valve motors became just as efficient at high speed and still had the benefits of a wide power band.

In some the intake port is open for a full 360 degrees and the reed valve completely controls the intake timing. As the valve only opens on demand – i.e. when the pressure in the crankcase is low enough – and closes whenever the crankcase pressure goes above atmospheric – and as it is easier to build a reed valve engine than a disc valve type, it is not hard to see why reed valves have become so popular.

Increasing the intake open-time and streamlining the port will allow greater gas flow and sustain the gas flow at higher engine speed.

Yamaha's work on reed valves eventually allowed improved high-speed performance as well as extending the rev range for which the valves were originally designed. They did this by establishing optimum sizes of reed block for intake port sizes and by choosing petal frequencies to match the desired peak engine speeds. They also used the chamber behind the reed valve to help scavenging, by connecting it to an extra ('seventh') scavenge port. As well as providing more gas flow to the cylinder, this also relieved the pressure on the back of the valve, making it open faster when the next cycle began.

Finally the length of the intake tract can be important as it is open to fairly strong pulses from the crankcase. If the port is smooth these pulses will be reflected back and forth along its length; the arrival of a high-pressure pulse just as the port opens or as the reed valve begins to lift will speed up the intake process. As pressure waves travel at the speed of sound in the gas, their time of arrival depends upon this speed and on the length of the tract.

Scavenge

The very early, post-war two-strokes were largely based on the pre-war DKW design with Schnürle, or loop, scavenging. In this, fresh gas is directed at the back wall of the cylinder and goes up in a loop, pushing out the burnt exhaust gas through the port in the front of the cylinder. Unlike earlier designs it allows a flat-topped piston (instead of the deflector type commonly used to prevent the new gas mixing with the old). Obviously the speed and direction of the new gas is critical if it is not to mix with the exhaust gas or to be lost out of the exhaust port.

These early two-strokes had very poor crankcase pumping and when people tried to tune engines like the Bantam or Villiers, they found they could get large improvements by working on the crankcase compression. As some of these engines had bob-weight flywheels and the crankcases were very large, it was an easy matter to reduce the volume, increase the compression and thus raise the delivery ratio (or pumping efficiency). The delivery ratio is proportional to the crankcase compression while the engine speed at which peak delivery occurs is inversely proportional to the volume of the crankcase. So small diameter flywheels, with full-circle webs and all spare volume plugged with alloy 'stuffers' really made big differences to the engines' pumping abilities.

Some tuners took this to extremes and built elaborate stuffers to fit up inside the piston skirt and generally fill all the remaining empty spaces. This time any increases were totally offset by the poor gas flow which these things caused; crankcase compression was right up, but the gas couldn't find the ports! Worse still, the big end, small end and piston are lubricated by oil mist carried in the intake gas and this cooling, lubricating stream is usually aimed at the big-end. Putting things in the way did very little to improve big-end life.

High crankcase compression got a bad name because of this and because the Japanese GP bikes of the 1960s also used it. They were famous for their ultra-high revving engines and extremely narrow power bands. Useful speeds of 17,000 to 18,000 rpm were not uncommon, along with the need for 10 or 12 gears. They used very small scavenge ports and high crankcase compression in order to get very high gas velocities out of the ports. This scavenged the cylinder too effectively and much of the new gas was lost through the exhaust. They also used very short stroke engines, with small cylinders – 50 cc twins, four cylinder 125s, etc. Thus the crankcases were very small – and would give peak delivery at very high engine revs. Despite the designers' aims and despite the crankcase volumes, high crankcase compression became associated with narrow power bands and when the next generation of machines appeared, it was fashionable to ignore crankcase compression altogether. Instead, designers and tuners went to work on the direction of the scavenge ports, improving the flow and aiming it for maximum effect. In other words, they found that a well-directed stream of gas would remove the old exhaust gases and do it with minimum dilution and minimum loss of fresh gas.

Having achieved this, it seems reasonable to suppose that the same could be done but with a better delivery ratio, improving the cylinder filling and raising the torque. An increase in crankcase compression produces this benefit, as long as nothing is done to obstruct gas flow in the crankcase.

So far, though, development has revolved around more and more devious scavenge ports. Some explanation of the numbered ports might be helpful.

The very early two-strokes simply had intake, scavenge and exhaust ports and were often called 'three port engines'. However when the Japanese started developing two-strokes, they used disc valve engines and found they could fit an extra scavenge port into the back wall of the cylinder, as there was no intake port there. This improved engine performance and they called it the third (scavenge) port.

Later, Yamaha added two auxiliary scavenge ports to their piston-ported twins, and called them five-port motors. What was presumably the 6th port was a boost port in the rear wall of the cylinder. Actually it was no more than a groove and it trapped a pocket of gas behind the piston skirt. Later this port was fed by a window in the piston and was meant to promote gas flow around the small end in order to cool it. Another variation was a very small port which opened *before* the other scavenge ports. This was meant to transmit cylinder pressure to the crankcase, raising pressure there just before the scavenge ports opened.

When reed valves were being developed, Yamaha noticed that the chamber behind the reed valve contained gas at crankcase pressure, which was in a dead space once the piston had closed the intake port. They opened a 7th port from this chamber into the cylinder and got a useful power increase. Others have tried interconnecting the reed valve chamber with the other scavenge ports.

7

Exhaust

The timing and size of the exhaust port are critical to peak power. It opens when the burnt gas is still expanding and the drop in pressure has to make way for the scavenge gases from the crankcase. The new gas can only fill the cylinder while the exhaust is open, so fresh gas is able to escape into the exhaust port. After the scavenge ports close, the exhaust remains open, this time long enough for pressure pulses in the exhaust system to push back fresh gas into the cylinder.

Because of this the height of exhaust port will only work efficiently over a narrow range of speeds. Yamaha have used a variable-height port, while Honda and Suzuki have both used variable chambers (see Figs. 38 to 40) in the exhaust in order to modify the pressure pulse activity and consequently increase the rev range. Increasing the exhaust port timing will shift peak power to a higher speed and cause a reduction in low speed power.

Exhaust system

Because of the way in which the exhaust and scavenge porting work, the exhaust system is also critical to engine performance. From its dimensions it has to reflect pressure pulses back to the engine, arranging for a low pressure pulse to arrive when the exhaust is opening and for a high pressure pulse to arrive just before it closes. Again the effect is only beneficial over a fairly narrow speed range.

Matching

Peak performance of all parts concerned with the flow of gas is dependent on engine speed. Therefore it is essential to match them all, to work in the same speed range – that is, if maximum power is the object. A better spread of power, but with a smaller peak, can be obtained by making a careful mismatch between certain parts. This is often carefully orchestrated on road bikes to get the most suitable characteristics; before the engine can be tuned it is important to recognise what the manufacturer has done.

Restrictions

As well as speed-related effects, gas flow is physical. A bigger hole will pass more gas. A smaller hole will raise the gas velocity. It is essential to determine where volume flow and gas speed are important and to locate any physical restrictions. Obviously if the gas cannot flow, then no amount of port work will give more power.

Heat exchanger

Fuel added to the air flow is burnt – as efficiently as possible, which generally means as quickly as possible. The heat from the burning fuel raises the pressure of the gas in the cylinder and this does the work. It also raises the temperature and it is important to distinguish temperature from heat; they are not the same thing. Temperature is a symptom of heat and most of the time it is a nuisance.

When the gas is being compressed (before ignition) its temperature is raised; it is raised even further when the plug fires. Once its temperature is higher than the surrounding metal, heat will flow from the gas to the metal (and the symptom of this is that the temperature of the metal increases). This is heat that could be doing work on the piston instead of warming up the surroundings, so the more heat loss, the less efficient the engine is.

Turbulence inside the cylinder can even out the fuel mixture and speed up combustion. But it also brings more of the gas into contact with the metal of the head and the cylinder providing the opportunity for more heat loss. In this respect, more turbulence will cause less power.

The design of the scavenge streams and of the head should cause enough turbulence for optimum combustion – and no more.

At high speed there is less time for heat loss and the efficiency (the adiabatic efficiency) goes up. Increasing the compression ratio and optimising the ignition (timing and spark duration) also help to extract more heat from the gas.

Fuel injection

Fuel injection does not offer an improvement in terms of air flow compared to carburettors – it suffers the same problem of needing a throttle valve which causes some obstruction and which increases the engine's pumping losses on part load. It has the added disadvantage (on a race bike) of needing a fuel pump, regulator, an electronic control, a stable power supply (which means a battery and a generator) plus suitable protection for all these units. There are circumstances in which fuel injection could increase power: (a) in installations where there is not enough space to fit a carburettor body, (b) where there is a heat soak problem which could cause fuel vapour locks in the carburettor float bowl or fuel line and (c) on engines with narrow power bands or sudden changes of torque, where it is very difficult to tune carburettors to meet steady speed and transient requirements.

If fuel is injected into the air flow before the crankcase then it suffers the same problems as carburettors, namely that obstructions such as reed valves and the crankshaft tend to encourage larger fuel particles to drop out of the air stream, collect on hot engine parts and then evaporate. This takes some time to stabilize and is one reason that engines need a rich mixture for acceleration and maximum power. The lighter fuel fractions evaporate first and as these can have a different octane rating to the heavier fractions, the octane value of the fuel reaching the cylinder can fluctuate.

Injecting fuel into the scavenge ports removes most of this problem but introduces others in the form of less time to mix the fuel and air, plus the need for two injectors per cylinder in order to distribute the fuel evenly. It still leaves the problem of fuel 'short-circuiting' (escaping into the open exhaust port and being lost) which is the reason that two-strokes have high hydrocarbon emissions in their exhausts. This can be avoided by injecting fuel after the exhaust port has closed (when fuel is sprayed into the cylinder it is called direct injection; adding fuel to the air flow at the scavenge ports or further upstream is known as indirect injection). The problem then is that there is even less time for adequate fuel/air mixing and the fuel system needs to operate at a higher pressure in order to overcome the rapidly increasing cylinder pressure.

One solution to this, developed by Lotus Engineering in England and separately by Orbital in Australia, is to inject fuel into a small chamber of air, pressurized by a separate compressor or by an air bleed from the main cylinder. This very rich fuel/air mixture is then introduced to the cylinder, via an injector nozzle and valve in the Orbital engine, via a rotary valve in the Lotus engine. There are two benefits: one is that the fuel is brought in to the top of the cylinder, near the spark plug and furthest from the exhaust port as the exhaust is closing, so the risk of short circuiting is enormously reduced; the second is that the method of mixing fuel and air produces what is known as a stratified charge.

A rich mixture is easier to ignite, in terms of spark plug voltage and the speed at which the flame travels through the mixture. Having a rich mixture near the spark plug and a lean mixture further away – a 'stratified' charge – means that combustion is easy to start and spreads quickly. The expansion and increase in temperature which this causes drives unburnt fuel towards the weak, outer regions and excites the remaining gas to a point where it will also burn quickly and completely. That is the theory. The proof is seen when an engine will deliver the same (or more) torque for less ignition advance; with better fuel consumption and with less unburnt or partially burnt fuel in the exhaust.

Trapping valve

Direct fuel injection may be able to clean up two-stroke exhausts but it doesn't prevent fresh air short circuiting through the exhaust. The problem is that to get sufficient blowdown to remove the burnt gas at high engine speeds, the exhaust has to open early and then, of course, it remains open an equal length of time after the scavenge ports have closed. A trapping valve is designed to close off the exhaust port (or the top half of it) after the scavenge ports have closed. Different versions have been described by Lotus and by Boyesen in America. In the Lotus design (Fig. 3) the valve (similar in appearance and position to the Yamaha power valve, Fig. 38 (a) is held fully up, level with the top of the exhaust port at high engine speeds. It is operated by an eccentric driven at engine speed and controlled by a variable linkage. After the piston has descended to BDC and is beginning to rise, the mechanism starts to lower the valve so that by the time the piston has risen far enough to close the scavenge ports, the valve is closing the remaining open portion of

Fig. 3. Layout of the trapping valve developed by Lotus. The eccentric shaft is driven at engine speed and closes the exhaust port when the piston is rising and has closed the scavenge ports. The position of the timing control lever affects the height to which the valve is lifted, effectively controlling the top edge of the exhaust port and therefore its timing. *(Lotus Engineering)*

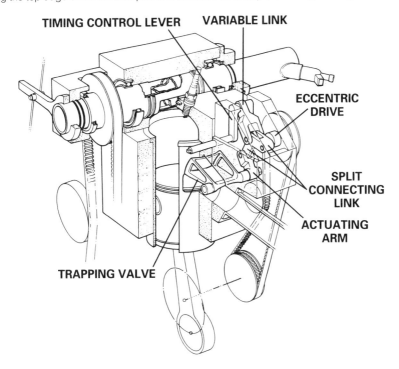

TIMING CONTROL LEVER   VARIABLE LINK

ECCENTRIC DRIVE

SPLIT CONNECTING LINK

ACTUATING ARM

TRAPPING VALVE

the exhaust port, preventing any loss of fresh gas. As the piston reaches TDC and begins to descend, the valve is fully opened. The valve now effectively forms the top edge of the exhaust port – like Yamaha's power valve (Fig. 38 (a)) – and the height to which it is raised is governed by the position of the operating linkage. At high engine speed it will reach the top of the port; at lower engine speeds the linkage can be moved so that the valve will be parked in a lower position, reducing the time-area of the port to match the engine speed. The action of the trapping valve tends to make resonant exhaust systems redundant, although low pressure in the exhaust pipe at the time the exhaust port opens would still be useful. Multicylinder engines could have plain header pipes collected into one secondary and one silencer, which would save on weight, space and cost.

Mechanical efficiency

The final tuning aid is to make the machine turn as efficiently as possible. This involves precise assembly; checking the truth of crankshafts and other critical pieces; setting piston clearances accurately and generally making sure that all frictional losses are kept to a minimum.

Chapter 2
Tools and equipment

The main object of development work is to get results; successful results, that is. Race teams are only interested in more power and better reliability, which is unfortunate because most of the time it is more important to discover why things happen – and this includes the things which reduce power as well as those which increase it.

Once the tuner understands the mechanism involved he can save a lot of time and trouble in the future. But running a series of pure experiments is a luxury which few can afford. The idea of an experiment is to test a theory and to produce results which are repeatable. In this respect the 'failures' can be more important than the successes – what matters is that you understand the theory and can apply it to improve any engine you build in the future.

This sort of a programme is costly in terms of time and of engines but the next best thing is to keep a detailed record of all work done, and the results obtained. First it provides an indisputable reference, so the same specification can be repeated accurately whenever it is needed. Second, if there is some method of cross-referencing items, it can be used to pick out patterns (for instance in exhaust system dimensions) so that you can predict a probable starting point and eventually make up your own formulae.

To this end, it is important to record your notes in the same format each time, to make them concise and easily readable and to develop some form of index. I'd suggest using an A4 size ring binder and notepad so that pages can be inserted in any order. Give each test a clear title and date (to avoid confusion if you work on a similar model later).

The best format is probably something like the specification tables the manufacturers use in their shop manuals – but only include the details which you've changed or which are of special interest. Follow this by all the test results and wind up with your own comments and conclusions.

A typical start would be to measure the engine – port dimensions, piston clearance, head volume and so on – and list this together with scale sketches of the port arrangements. A pad of graph paper is useful for this, and for plotting test results in an easily-comparable form.

The next step is to calculate the state of tune of the engine, for example by working out the time-area values for the ports (see Chapter 4 and Appendix). This should indicate whether the various components are 'matched' closely enough or not and from here you should be able to work out the first steps of a development programme. Methods of assessing engines in this way are covered in Chapter 3.

Checking the power output and fuel flow obviously requires equipment like a dynamometer and flow meters, and the sad fact is that there is no substitute for measuring output under controlled conditions. I know one or two tuners who have the patience and the sensitivity to set up engines on the track but they are the exceptions and, even with their special aptitudes, it is a very long process, subject to all sorts of changes from the weather, the machinery and the rider. But, if there is no access to a dynamometer, some other means of testing will have to be found. A track can be used, provided you're aware of its limitations – and if you can sit around waiting for the right sort of weather. Examples of track-test methods are shown in Chapter 10.

An alternative way of filing all the data away is to use a computer with a disc filing system. Given suitable fields, this could produce all the data for one machine or could compare items (e.g. exhaust timing and peak revs) for a number of different engines.

Measuring tools are a necessity, and the tuner will find that there are varying degrees of accuracy while all require a fair amount of expertise and practice before they'll give consistent results.

Feeler gauge (thickness gauge)   Used in combination they can usually be arranged to measure to 0.04 mm. The accuracy is dependent on the user, especially when adjusting a gap down on to a feeler gauge when it is necessary to judge how much to grip or pinch the gauge. Usually, when checking a plug gap, piston ring gap or bearing endfloat, there will be closely defined limits and the gauge can be used as a go/no go tool, e.g. if the gap is 0.6 to 0.7 mm, then a 0.6 mm feeler should pass through it, while a 0.7 mm feeler should not.

Caliper   Available with legs for internal or external measurement; inexpensive and useful for comparing port widths, etc., or for measuring

Fig. 4. Vernier calipers

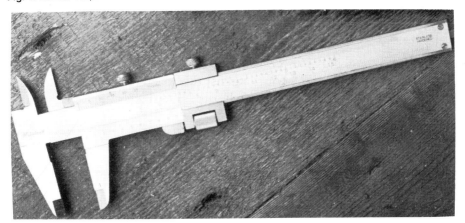

inaccessible things like port windows when the measurement can be taken from the caliper using a vernier caliper.

Vernier caliper   The sliding vernier scale will measure to 0.02 mm. The ground jaws will take internal and external measurements and the instrument usually includes a depth gauge, which is useful for measuring port heights. One of the easiest types has screw clamps to lock the sliding cursor and a knurled wheel so that small changes can be made precisely.

Dial gauge   With suitable clamps or adaptors to hold it in position, the dial gauge is one of the easiest measuring instruments to read. It can, however, be tricky to set up as there must be no rock or wobble in its mounting and the shaft must be in line with, or square to, the object being measured. Usually the gauge reads to 0.01 mm. Its main uses are to check shafts for runout and endfloat (in conjunction with V-blocks and a surface plate where necessary) and to set the piston in the exact position for ignition timing, or locate TDC.

Micrometer   The most accurate of the measuring instruments, this is usually calibrated to 0.01 mm. The moving thimble can be turned via a ratchet which ensures an equal pressure when the micrometer is tightened on to the object to be measured. A standard test piece is supplied so that the instrument's basic accuracy can be checked. The usual sizes are 0 to 25 mm,

Fig. 5. A dial gauge, with an adapter to screw into the spark plug hole

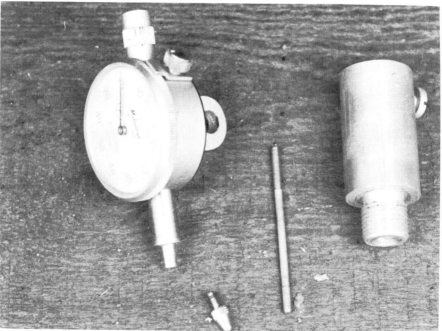

Fig. 6. Micrometer – for taking accurate external dimensions. Internal micrometers are available, or the external sort may be used in conjunction with a bore gauge

25 to 50 mm, etc. For internal measurements, internal micrometers are available or an external type can be used in conjunction with a bore gauge. Essential for accurate measurement of shaft/bearing diameters, piston diameter, bore diameter, etc. To avoid errors caused by thermal expansion, parts must always be measured at the same temperature. This is very important when measuring two components, e.g. a cylinder bore and piston, in order to work out the clearance between them.

Degree disc Extremely useful, as long as there is a rigid, concentric mounting for it. Indicates crank position to check port timing, piston height, etc. Don't rely on it for ignition timing.

Straight edge Usually a steel rule or steel try-square; useful reference for testing truth of gasket faces, etc.

Surface plate Accurately flat, heavy steel plate, used as a reference for making measurements (often in conjunction with V-blocks) – for checking connecting-rod dimensions and truth, crankshaft truth, etc. Can also be used for checking for warp in machined faces on large castings.

V-blocks Accurately machined steel blocks with large V-notches to support round shafts.

Scriber Pointer with sharp, hardened steel or carbide tips, used for marking metal to be machined.

Stroboscope Used for checking ignition timing and advance range while the engine is running.

Multi-meter For checking electrical circuits for continuity, etc. Useful for setting precise timing on contact-breaker ignition systems. The slightly more expensive type which reads resistance (the cheapest ones only have a kilo-ohm scale) can be used for testing electronic ignition circuits, pulser coils and so on.

Burette For measuring volume of liquid. Used for checking combustion chamber volumes, etc.

Being able to measure and mark-up the engine, all that remains is to be able to modify it. There are a few calculations which can speed up the inevitable trial and error process and they can be made a lot less tedious by the use of a calculator, preferably a programmable type, or by using a computer. Sample programmes for both are given in the Appendix.

For basic engine modifications, a simple selection of files is about all that is needed, but at various times you will need access to some machine tools. In particular a hone (and possibly a boring bar), a mill, a lathe and welding equipment for thin steel sheet and for aluminium alloy. Fortunately most of the jobs are straightforward and most machine shops will be able to do them accurately and cheaply. As exhaust pipe work involves a lot of chopping and welding, this is one art which is worth learning.

For porting work you will need some small flat and half-round files; sharp, second-cut types are probably the best – the soft alloy quickly clogs any fine file. If possible it is better to use a rotary file (a variety of shapes are available) along with a flexible drive.

This can be powered by an electric drill clamped in a bench stand, or by a special motor. The problem is that the cutters work better in the alloy when they are turned at high speed.

It is important to keep the flexible drive as straight as possible. Many tuners have the motor suspended very flexibly, so the whole assembly can move with the actions of the operator and this way they can use high speeds and still get a reasonable life out of the drive. The best power tool is the type – either pneumatic or electric – designed to turn at speeds from 15 000 rev/min to 25 000 rev/min. These make porting work very easy, do not destroy flexible drives and do not clog up the bits. Unfortunately they are very expensive.

The actual shaping of the port window (in the cast iron liner) should be finished with small needle files, or, to get into awkward corners, with curved riffler files. These come in all imaginable shapes and sections.

One of the problems with porting is that it is difficult to see what you are measuring, particularly when the edges of the port windows are chamfered and they follow curves instead of straight lines. Working on a bench with a light-coloured surface and using an Angle-Poise style of lamp will help to get some light into the barrel and the ports so you can at least see the outline clearly. After that it is a matter of taking a lot of care and, if there's any doubt, repeating the measurement several times until you get consistent answers.

Once the port outline has been made to give the required timing and area, the rest of the port must be blended in to suit. At one end of the port is the window and at the other there is the exhaust pipe, reed block, etc. The transition from one section to the other must be as smooth as possible. Steps, ridges or sudden changes in section will disturb the pressure pulses and take energy from them, in exactly the same way that baffles remove energy from the exhaust gas in order to silence the engine. The diagrams show the right and wrong ways to modify the port shapes.

When it is necessary to prevent pressure pulses from influencing the gas flow, the best way to do it is to use a sudden step. In extreme cases, something like a washer, fixed into the entry to the exhaust pipe, is used.

Fig. 7. There are several ways of altering a port's dimensions – the three marked 'bad' would not work. Ragged casting material should also be removed to make the lines of the port as smooth as possible

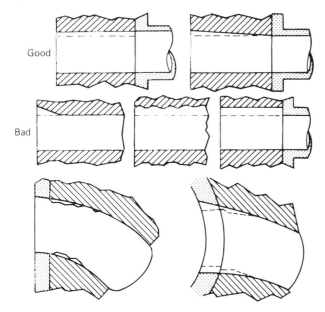

Fig. 8. Port window edges should be chamfered as shown to minimise wear on the rings. Corners should also be radiussed and a curved window is less severe than one with straight edges. A bridge across a port will run at a higher temperature and should be ground to a radius as shown to prevent it expanding and rubbing on the piston

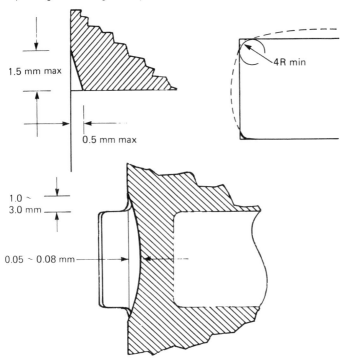

Fig. 9. Pistons and cylinders should be measured in these positions. The bore is measured on the axis of the piston pin and at right angles to it, at X – 15 mm down from the top of the cylinder; at Y – just below the port windows; and at Z, on the unworn, lower portion. The difference between the maximum and minimum measurements gives the liner's taper or ovality. Pistons are usually measured at A – at right angles to the pin and just below it, or about 15 mm up from the bottom of the skirt. The piston clearance is A subtracted from the liner bore at X, at right angles to the axis of the pin

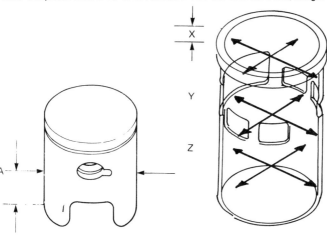

While intake and exhaust ports need to be smooth to promote gas flow and to allow pressure waves unrestricted travel, the scavenge ports play a slightly different role. Here the gas is being pumped through them and the important factor is the direction in which each port aims its gas stream.

First, the ports must make a symmetrical loop, with all streams converging on the cylinder head and this usually means that each pair of ports must open simultaneously and direct their own stream of gas at equal angles across the piston crown. Second, the upward angle of each stream should be the same as its partner. This can be checked by flowing water through the ports, or by using an air flow rig with marker dust or smoke injected at the port.

Generally you would want to avoid turbulence in the scavenge ports and this is done by making the port a nozzle, starting with a wide entry and gradually tapering down to its smallest section. This can usually be combined with the tightest part of the port's curving path, so that the gas stream makes the smoothest possible flow. Simply opening out the parts of the port which are easy to reach will make the port shape about as bad as it is possible to make it, so unless you've got the equipment to get to each part of the port wall, you will probably be better off leaving the scavenge ports alone. Flexible moulding material is a useful way of checking on scavenge port shapes, and will show how similar each port is to its partner.

The final consideration for port windows is to put a radius into each corner – the larger the better as far as gas flow is concerned – and after final cylinder honing, the port windows should be chamfered to give the piston ring the longest possible life. Recommendations on this vary, some manufacturers, like Honda, using a radius of 2 mm on horizontal edges and 0.5 mm on vertical edges. Others, like Suzuki, specify a chamfer of 1.5 mm × 0.5 mm, as shown in the diagram.

Piston clearance is not easy to determine, mainly because pistons are not round, they are slightly cammed and their sides are not parallel, they taper towards the crown. The manufacturer specifies his own method of measuring – usually the piston is measured just below the piston pin, on an axis at right angles to the pin. The only snag is that on some engines the back skirt of the piston is cut away. The barrel is usually measured 15 or 20 mm down from the top, on the front-to-rear axis. The official piston clearance is the difference between the two measurements.

Now, if the piston has been modified by taking large pieces out of the rear skirt, it will have to be measured somewhere else.

The only solution is to measure the piston as close to the original position as possible, keeping at 90 degrees to the pin. Measure an unworn piston at the same point and use this value to obtain a new piston clearance dimension, to be used in place of the official figure. (See piston clearance, Chapter 6.)

Chapter 3
Engine evaluation

Assuming that there is some choice in the matter, it always helps to be able to use the best engine for the job. Ideally this would be one which was so far ahead of its competitors that you wouldn't have to do anything to it. That is, until someone else got one.

The motives for tuning the engine have to be considered – if it is for competition use, then pure horsepower is the only criterion. This is an entirely different matter to tuning an underpowered road bike to keep it on a par with current models. The first question is about the engine's starting point; what is its level of performance compared to its competitors and its cost? Would it not be easier to change the entire engine?

Usually we are stuck with what we've got, so the second question is about the engine's adaptability. Is there a bigger engine in the same range which shares the same parts? Can it be converted to a larger displacement? Are there any performance/competition parts available for it? Is there a competition version, e.g. a motocrosser?

If the answer to any of these is 'yes', then it will probably be fairly easy to uprate the engine – how far this can go will depend on the likely reliability. So once again it is back to the shop manuals and parts lists. Is the lubrication system the same as on the larger model? What about crankshaft bearings, clutch and gearbox bearings? If it shares any of these structural components with a *smaller* version, then reliability may become a problem, at least if the standard parts are used.

Finally, are there any unusual features which may be difficult to modify or extend? One likely problem here is the machine with a plated bore, whether it is chrome, Nikasil or some electro-fusing process. It may mean that the cylinder cannot be rebored and that it will be difficult to modify the port windows without damaging the plating material. Mahle can recondition Nikasil bores as long as the base metal is not damaged.

It is sometimes possible to establish a starting point, and to estimate how much work there is to do, by comparing the broad specification of the engine with state-of-the-art machines.

Some care is needed here because different bore and stroke proportions can disguise the state of tune of the engine. For the same piston displacement long-stroke motors have considerably more cylinder wall area than 'square' or short-stroke motors. There is consequently much more room for port windows and it is possible to build in considerable port area, without going to a very high state of tune. The advantage is illusory because the long-stroke motor loses out in so many other ways. A comparison of three motors (long-stroke, square and short-stroke) shows the differences.

	long-stroke	square	short-stroke
bore × stroke, mm	52 × 58	54 × 54	56 × 50
displacement, cm^3	123.2	123.7	123.2
cyl wall area, cm^2	9470	9156	8792
exh opens, deg ATDC	90	90	90
exh port height, mm	25.3	23.6	22.2
exh port width, mm	36	37	39
port width, % of bore	69	68.5	69.6
port area, mm^2	910.8	873.2	865.8
port time-area, s-mm^2 at 8000 rev/min	2.24	2.15	2.12
piston area, mm^2	2123.7	2290	2463
mean piston speed, at 8000 rev/min, m/s	15.5	14.4	13.3
max piston accn at 8000 rev/min, m/s^2	25415	23662	21910

To go through this table, a line at a time, we have the three motors; the 52 × 58 long-stroke is the traditional post-war utility two-stroke, taken from the DKW design. The 125 BSA Bantam is the classic example. Next there is the 54 × 54 motor which was, for a long time, the definitive sports 125 cylinder layout. Finally the short-stroke 56 × 50 motor which was always the alternative to the square layout and finally succeeded it for high speed, high output engines.

All three give displacement in the 125 category and have been used as singles and twins. Keeping the same stroke and increasing the bore to make 150 and 175 cc cylinders showed the potential of the big-bore/short-stroke engine.

The cylinder wall area shows just how much room the long-stroke gives for its exhaust and scavenge ports; the bad news is that this is all extra metal to which the gas can lose heat. The long-stroke inevitably gives the scavenge gas a longer path to travel – which takes valuable time.

All of the exhaust ports have been allowed to open at 90 degrees after TDC, to give the engines the same nominal state of tune; in fact it is quite a low state of tune for the short-stroke and a very high state for the long-stroke; a tuned Bantam of the early 1970s would probably open its exhaust about 97 degrees after TDC.

Exhaust port width is the first place that the long-stroke suffers. We've allowed it 36 mm, which at nearly 70 per cent of the bore is the practical maximum currently allowed by piston rings; a contemporary engine would have had between 32 and 35 mm port width. With the same degree of ring reliability the short-stroke motor can use a port 39 mm wide.

Even so, the long-stroke makes up on the total port area, with an increase of just over 5 per cent on the short-stroke motor – implying that the long-stroke could still flow more gas. This appears to be confirmed by the port's time-area – a calculation which describes the *amount* the port is open multiplied by the *time* which it is open at any given engine speed (this is explained fully in Chapter 4, and the calculations are given in the Appendix). Calculations based on real long-stroke engines give a time-area of around 1.6 to 1.7 s-mm^2 at 8000 rpm. However, given that the engine could reach the state of tune shown in the table, we are looking at a port area increase of 5 to 6 per cent in favour of the long stroke.

But power comes ultimately from gas pressing on the piston crown and the short-stroke motor has almost 16 per cent more area for it to push against. That, given the same gas pressure, would result in a force some 16 per cent higher. The answer to this is that the long-stroke motor has a longer crank (16 per cent longer) and therefore exerts a greater torque on its crankshaft, so things *should* even themselves out, but from here on things start to work in favour of the short stroke.

The long-stroke means that the piston has further to travel, and so, at any given crank speed, the piston in the long-stroke has to travel faster. On the expansion stroke, the gas in the cylinder is trying to press down on a piston which is rushing away from it – at an average 15.5 m/s, compared to the 13.3 m/s of the short stroke when both engines are turning at 8000 rev/min. This also results in greater friction and more mechanical losses than the short-stroke engine.

The final figure, which is the piston acceleration when it reaches a maximum at TDC, shows the significantly lower stress caused by the short-stroke layout (the calculation for this is shown in the Appendix; for the same piston mass it represents the maximum force generated in the connecting rod when the engine swings through TDC). The short-stroke motor can tolerate much higher engine speeds before it reaches the same level of stress as the long-stroke version.

With the benefit of hindsight and other people's experience it is easy enough to point out the advantages of the short-stroke but the theoretical figures put on its port timing and time-area do not immediately suggest that it may have an advantage over the other two. It is not surprising that manufacturers persevered with long-stroke designs, especially when materials and lubrication did not permit very high crankshaft speeds and at that time there was no way of knowing that a short-stroke motor would tolerate such extreme port timing. It also explains why the 54 × 54 motor has been used so extensively – it manages to achieve the best of both worlds in most respects.

The moral is that it is important to make realistic comparisons and not to assume, for example, that any engine can be given port timings comparable with a current GP racer. But using a spec like the one above, plus items like carburettor size, compression ratio and peak engine speed, will show where an engine has deficiencies – or advantages – over another.

Many roadster engines have built-in restrictions which often do not show up clearly on such specifications (small carburettor size or abnormally low peak speed are the exceptions). The restrictions are often to make the engine comply with a particular set of legal requirements – either a power or a speed limit. Alternatively, there may be more subtle restrictions built in by the manufacturer to make the engine more flexible, easier to ride and more economical. It is important to recognise these restrictions before attempting any other modifications. Usually they take the form of restricted intake passages, exhaust restrictions or steps in the exhaust/intake to dissipate the energy in pressure waves. In one case it was nothing more complicated than a stop to prevent the carburettor throttle reaching full lift. Occasionally a manufacturer builds an engine which is restricted all the way through – the carburettor is small and all the ports match its size; the exhaust is restrictive and even the ignition may have some rev-limiting device built it. Obviously it is not worth the trouble and expense involved in trying to tune such an engine.

Development work also has to meet certain limitations; in competition there are capacity classes plus restrictions on certain parts in production and F1/F2 type events; racers have to meet certain noise limits and there are rules governing the type of fuel which can be used. Street bikes have more stringent regulations, from the EEC1015 noise test through to various capacity and power limits governing unqualified riders.

Development programme

Having had a good look at the engine and worked out likely modifications, the next step is to devise a logical sequence of tests which will give the maximum amount of improvement for the minimum amount of effort. The word 'improvement' has to take on a highly subjective meaning here because you are only likely to see gains in certain areas.

Set a target performance. Depending on what you hope to achieve with the machine and on how suitable the engine is for further development, there are several choices open.

1 Torque

To raise the power output without changing the engine speed means that the engine torque has to be increased. This wouldn't be a bad idea if the engine already had a heavy piston or the piston speed was close to the practical limit. One way to achieve this is to increase the air flow through the engine – by increasing the bore size, by reducing any restrictions to the air flow and by matching the port timings to the desired speed range. Another way is to extract heat more efficiently from the gas, by raising the compression, paying close attention to piston clearances and the squish clearance at the cylinder head, and by careful fine tuning of the ignition and carburation.

With this type of work you are not changing the nature of the power curve, just trying to get more of it. All you can do is try to improve on what the factory have already done, making the difference between a production-line engine and one that has been hand-built. The changes are likely to be small.

An extension to this line of development is to improve bottom end torque, at the expense of peak power, for example to make a trials engine. Here the requirement might be to shift peak torque further down the rev range; equally this sort of engine needs reliable low-speed running and an excellent pick-up, with the best possible throttle response. Raising the compression will help, and the porting can be arranged to be most effective at low speed. Work with reed valve and exhaust dimensions can change the power characteristics quite dramatically but to get the required response it will be necessary to put in a lot of work with ignition settings and part-throttle carburation.

2 Crank speed

Raising the crank speed will, if the torque doesn't fall too much, also raise the power. This can be a useful way of setting a target – if the original motor has peak torque at 8000 rev/min, the target could be to maintain this air-flow at a speed of 9000 rev/min, an increase of 12.5 per cent. This increase immediately tells you how much more port time-area is needed and what order of change must be made at the exhaust system.

Changing peak speed, even if the torque does fall, can be used to change the engine characteristics. It can be used to shift peak torque closer to peak power – or further from it. There may not be an increase in peak power but there could well be an increase either side of it, where the power no longer falls off so rapidly. This could make the machine more flexible, or easier to gear for some circuits. Being able to let it rev out can mean that you don't have to change gear in awkward corners; it may also give the opportunity of slipstreaming faster machines if your engine doesn't run up against a stone wall as soon as the power peaks.

These characteristics can often be tailored by changes in the exhaust system and possibly in the jetting and ignition timing. It isn't unknown for people to have a different exhaust to suit different circuits, or even different weather conditions.

3 Power

Raising the torque or the speed while holding the other constant will increase the horsepower, but to get real power changes it is necessary to do both and this means radical alterations all the way through the engine, plus predictable losses in mid-range power and reliability.

Because of the need to match the intake, porting and the exhaust, it is difficult to approach this level of tune in single, small steps. One way to dodge this all-or-nothing approach is to use two or three barrels and pistons, leap-frogging the modifications from one to the next so that only the successful changes are incorporated and, when you go too far, it is fairly simple to revert to the previous set-up.

Once a power target has been set, the development programme can start. An outline of typical steps – and problems – follows:

It is often possible to make a few quick/cheap experiments with the machine to see which systems it responds to and to try to find out where it is restricted. Depending on what is available, the sort of alterations which can be made are to the silencer and air box, changing the exhaust, experiments with any variable exhaust valving or chambers which are connected to the intake or exhaust. Quite often these parts, particularly the air box, produce bigger changes in carburation than they do in power, which is something you need to be prepared for. The engine's sensitivity to port timing can *sometimes* be seen by raising the barrel, using extra base gaskets to lift it 0.5 to 1.0 mm. This has the effect of raising the exhaust port and the scavenge ports – so the exhaust blowdown period isn't changed. It also lowers the compression ratio and, depending on the design, it may change the intake timing as well. Also, because it raises the floors of the exhaust and scavenge ports, as well as their ceilings, it doesn't show the effect of increasing their areas. Despite the fact that it has become a popular preliminary modification, it changes so many things (most of them for the worst) that the results are usually quite meaningless. If the engine needs an increase in both exhaust and scavenge periods, this test will confirm it; otherwise it is likely to be inconclusive.

For a slight amount of tuning, or to tailor the power characteristics, work on the exhaust system and intake will probably be enough, coupled with final, fine carburettor tuning. Any further work will require port changes, plus alterations suggested by the preliminary experiments, so the development programme should start here.

The first step is to take the engine out to its final size, either by boring or by fitting a new cylinder. If this is a conversion between two models (e.g. 50 cc to 80 cc, or 100 cc to 125 cc), check also that any changes in the lubrication system, clutch or bearing specifications have also been made. Remove any obvious restrictions, like throttle stops, washers in exhaust headers, big steps in exhaust or intake ports, etc. Unless the carburettor is ridiculously small, leave it for the time being, and also leave the air box and filter, as they are not likely to be restrictive on small engines. The carburation will have been set to work with the airbox and the advantages in the early stages of having clean carburation more than outweigh any slight increase in air flow.

To increase power or raise peak revs, calculate the necessary change in exhaust port and work towards it in two or three steps. If the engine becomes too peaky, modify the scavenge porting to increase the midrange. (See Chapter 4.) At each step it may be necessary to tailor the exhaust and to re-jet the carburettor. Finally there will be a mismatch between intake and exhaust, as the exhaust tries to push peak revs higher yet the intake cannot flow enough air at this speed.

At this point the intake time-area will have to be extended, or the restriction removed, depending on whether it is at the air box, in the carb itself or in the reed block. The changes in carburation will often give the answer, because an air flow restriction which is upstream of the carburettor will cause massive richness. When the carb itself is the culprit, it usually starts to work at its best, because the airspeed through it is so high, and there is a tendency for it to run rich at high speed on jetting which is correct at medium speeds. Poor atomisation may suggest poor airflow, perhaps bad pulse effects or a restriction downstream of the carb.

Once the porting has been successfully matched, the compression and squish clearances can be raised towards the practical limit, leaving only the fine tuning to be carried out at the end. Obviously enough carb and ignition adjustments have to be made during development to keep the engine from damaging itself. As the load and speed are increased it will probably be necessary to uprate the lubrication and possibly the cooling system as well.

In any development programme there will be mechanical failures and it is important to find the cause so that it can be cured or avoided in the future. Typical problem areas are:

1   Piston failure

- holed crown (plug too hot, ignition too far advanced, mixture weak or detonating)
- melted edge or top ring land (combustion temperature too high, excessive back-pressure in exhaust, detonation, insufficient cooling, incorrect piston clearance)
- seized (or smeared ring land) (running too hot, insufficient lubrication, poor piston material, wrong piston clearance, running at high speed or under load before the engine has warmed up)

2   Ring failure

- ring peg loosened (faulty manufacture; if peg is near exhaust port it may be caused by the ring expanding into the port and the ring end fretting against the peg when the ring is pushed back into its groove. Relocate the peg near the rear of cylinder if possible)
- excessive port width, port windows not chamfered (correct windows or renew rings more frequently)
- incorrect groove clearance (allowing ring flutter)
- incorrect type of ring (too thick or too heavy, will flutter at high speed)
- excessive piston/bore clearance or excessive cutaway in piston skirt, allowing piston to rock (may cause ring failure or damage the ring lands and make ring stick)

- ring trapped in groove – caused by piston touching head and squashing the top ring land.

3   Small end
- seizure or broken roller bearing – insufficient lubrication or cooling (boost port through piston skirt may help)
- piston running too hot/insufficient clearance between pin and piston
- incorrect assembly (circlips not fitted correctly, wrong thrust washers – if any – on small end)
- piston too heavy/speed too high/bearing and pin not strong enough for inertia loads.

4   Crankshaft and big end bearings
- usually speed-related failure of the big-end cage. May be caused by overheating or insufficient lubrication or by rollers skidding and increasing their working clearance. There are also problems with crankshafts twisting. Possible solutions: better lubrication, lower crank speeds; better quality bearings; better assembly (alignment, tighter press fit); more frequent overhauls. On a GP racer the crankshaft will typically be overhauled at intervals in the order of 200 to 500 miles, implying that it is essential to keep a careful log of the engine's running time.

Chapter 4
Air flow

The flow of gas into and out of the engine is the most important factor in regulating its performance; several properties of gases are brought into use and a brief summary of these phenomena will help to explain some of the techniques used in modifying engines.

First, gases have various energy levels, such as kinetic energy, which is proportional to the speed of the gas, squared. Then there is pressure energy, which is its pressure divided by its density, and heat energy – a function of its temperature and its specific heat. As long as energy isn't added to or subtracted from the gas, its *total* energy will remain the same but the constituent parts can be interchanged.

For example, if the gas is travelling along a pipe it will have kinetic energy. If it now arrives in a chamber, like a crankcase, and virtually comes to a standstill, this kinetic energy is reduced to zero, but the overall total has to remain constant as nothing has been taken from the gas. The result is that one or more of the other energies will rise; the usual effect of letting gas rush into a static chamber is that the pressure rises. Exactly the same phenomenon is used in carburettors, where the air is drawn through the venturi at high speed. This change, from the virtually static air in the atmosphere around the carb means that the pressure in the venturi is lowered and petrol, at atmospheric pressure in the float bowl underneath the venturi, is forced through the jets to mix with the air stream.

Because the pressure, volume and density can all be varied so easily, the only thing that remains constant is the mass of the gas, so it is usual to refer to the mass flow of air through an engine (speed or volume flow mean nothing – the speed in the carburettor is very high, inside the crankcase it is very low. Similarly, the engine takes a volume of about 125 cc and promptly compresses it into a volume of about 10 cc, so volume flow doesn't mean a lot, either. But the same quantity of gas has the same mass, whether it is compressed or expanded, standing still or whistling through a carburettor).

Rapid expansions set up violent pressure waves which travel through the gas – the exhaust noise is an obvious example – and these pulses can be put to some use. When a pressure wave meets a solid wall, it is reflected directly back and travels along in the reverse direction, still maintaining its original pressure energy. But when the same wave reaches an open end of a pipe, the pressure is released suddenly and a wave of very low pressure is reflected back along the pipe. A wave of low pressure is also reflected from an open pipe, returning along the pipe as high pressure.

A series of pulses travelling back and forth will take a certain time to reach each end of the pipe, depending on how quickly they travel and the length of the pipe. They travel at the speed of sound, which is beyond our control, but the length of pipe isn't and this can be arranged, within reason, so that the pulse emerges at a time which suits us.

If the pulse arrives just in time to be strengthened by another pulse from the engine, the pulse energy will accumulate and will rapidly become quite powerful. Obviously the arrival of the reflected pulses would have to coincide with the frequency of the pulses generated by the engine for this to happen, and it will only occur at one engine speed. The effect is called resonance. There can be harmonics, as well – for instance at half the resonant speed there will be a condition in which every other pulse is reinforced. It follows that there will also be intermediate speeds where the generated pulse is totally out of phase with the reflected pulse.

The pulses are generated by the sudden opening and closing of the ports, as the rotary disc or the piston skirt flashes across them. These actions are so sudden that the pulses can be very strong, but it also means that the full area of the port is not open to the gas flow. There is a computer graphics program in the Appendix which animates the action of the piston but, to imagine it in slow motion, uncovering a port, it first reveals a thin slot, which is enlarged as the piston travels further, becoming a deeper rectangle.

The port reaches its full size only momentarily as the crank swings through the dead centre position and the piston briefly halts and then begins its return journey, progressively closing the port.

The more slowly this happens, the more gas can flow through a given port, but if we increase the engine speed, we reduce the time available for the gas to pass through. So the effective port area, as far as the gas is concerned, depends as much on the time it spends open as on its size; and as its size is continuously changing, this is not an easy sum to calculate. The concept is called time-area and it simply means that a port of 1 mm^2 which is open for 2 seconds can flow as much gas as a port which has 2 mm^2 open for 1 second, or 10 mm^2 for 0.2 seconds. If you halve the time, then you halve the port's capacity as surely as if you blocked half of it up.

In the past it has been necessary to rely on approximations for port time-areas, but it is the sort of sum which computers are particularly good at and the Appendix contains the necessary calculations along with a BASIC program which will deliver the answers in a matter of seconds.

There are many ways in which the gas flow can be influenced and the easiest way to study them is to follow the gas as it moves through the engine.

1 Intake silencer

A standard roadster will usually have some form of silencer on the entry to the air box. It could be a series of baffled compartments but it is more likely to be a moulded rubber duct, with a tapering entry and webs cast along its passage. It is there to prevent pulses reflecting from the open end of the air box and to cut down intake roar when the throttle is opened. It can be restrictive in two ways, first as a physical obstruction to air flow and second, by preventing the intake system from making use of resonance. In this respect it may also be beneficial, because it will also eliminate the unwanted, out-of-phase effects as well.

Fig. 10. Typical intake silencer

Removing the silencer will often give a slight boost in peak power, with a drop either side; replacing it will give a wider spread, without the sharp peak.

Apart from any resonant effects, the intake silencer is often too small to flow enough air, especially when the engine has been uprated. This can be tested by enlarging the entry to the air box (assuming the air box is going to be retained on the final version of the engine) or by making other holes. A restriction here will, of course, increase the depression on the carburettor and will make the mixture rich; removing the restriction will make the mixture suddenly run weaker.

If the silencer does prove to be restrictive but the bike is to be used on the road, then it would be worth enlarging the entry to the air box until it will accept an intake silencer from a larger, more powerful machine. On acceleration the silencer probably lowers the overall noise level by 1 to 2 dBA.

2 Air box

This does two jobs on stock bikes – it provides a home for the air cleaner element and it acts as a large air reservoir for the engine. Sometimes gearbox breathers are fed into it, so that any oil vapour is recycled through the engine. The entry (with or without the intake silencer mentioned above) is sometimes restrictive and the air flow can be increased by making the entry larger, or by making additional entry holes.

On very powerful engines – say over 50 bhp – the air box could become restrictive. It isn't likely to be so on smaller, less powerful machines, but if the engine has been radically tuned, then the stock air box could prove to be too small. Even on high-powered machines, the air box has an advantage. It acts as a surge tank and usually gives a significant power increase all through the midrange, even though it may crop a slight amount of power from the peak figure. In most cases, the extended rev range is well worth the slight loss. On small engines it may not make any difference at all to the power output. For it to work as a still air box, it requires a certain, minimum volume, from a broad comparison of standard components, this would seem to be in the region of ten times the engine displacement, or roughly 1 litre for every 15 bhp, whichever is the greater.

However, on nearly all engines it will make a large difference to the carburation and this may prove difficult to correct. The reason is that at low air flow the air box will not make much – if any – difference to the mixture strength. As the air flow rises it will have a greater effect. When the box is removed, the jetting has to be changed so that it is made richer at high speeds, but not altered at low- to middle-speed.

This usually involves changing the main jet and the air jet (see Chapter 8)

Fig. 11. The air box is an important feature for mid range power

as the main jet which is right for high speed will then prove to be too rich at lower speeds. If the carburettor float chamber vents and the air jets are also fed from the air box, this could complicate things still further.

3 Air cleaner

The stock arrangement will be to have a large element housed inside the air box. Usually the element is not restrictive on smaller engines. It would become progressively more restrictive as the air flow went up, and would create more of a pressure drop at the carburettor(s). Unless there is evidence that the cleaner or the air box are restricting air flow then any road bike is probably better with them than without them. It is an essential piece of equipment on any dirt bike.

An alternative is to use individual, 'high flow' filters on each carburettor. This immediately loses the surge tank effect of the air box and, as they offer very little resistance to air flow, some fairly serious carburettor tuning will be needed. There are types which can be oiled and others which run dry, but because of their exposed nature they are prone to dirt clogging and to rain water contamination.

Some types also have a flat plate end cap, which, as it is fitted close to the carburettor entry, can reflect intake pulses quite strongly. Occasionally this upsets the carburation quite seriously and the motor will not respond to normal jetting changes. If this is caused by the endplate, then it can be cured by making an adaptor to lengthen the intake tract – the easiest way being to mount the filter itself on a longer stub.

4 Carburettor intake

The open end of the carburettor will reflect intake pulses back towards the engine and this can be useful on all types of two-strokes. It can be used with an open carburettor, or fitted inside the duct to the air box; and possibly, if the filter were large enough, with an individual filter.

The beneficial effects happen when a high pressure pulse returns towards the engine just as the intake opens. This would help the fresh air move into the engine or, on a reed valve motor, would start to lift the reed petals, causing less obstruction to the air flow. The same high pressure pulse, reflected back from the reed block or piston skirt would next be returned as a low pressure pulse – which would make the reed valve close more quickly.

The bad effects are that gas blows back out of the carburettor, forming a fuel spray which can reach almost as far back as the wheel. This would mostly get lost when the bike was being ridden, but it is clearly visible when an engine is run on a dyno. The fuel spray can be prevented by altering the intake length – usually by fitting a different-sized bellmouth or trumpet to the carburettor.

There are various formulae quoted for the optimum length, and for the curvature of the entry, but they all have to make assumptions about the speed of the pulses (which also involves the gas speed, which is not constant anyway, and the speed of sound in the gas, which depends on its temperature and density) and in the end, all it can do is give you a starting point. What you do then is what you can do anyway: make it longer or shorter and see what happens.

There is no doubt that the size and shape of the intake have a significant effect on both power production and the behaviour of the fuel spray. The smoother the passageway, the more pulse energy will be carried along; sudden steps and changes of section will dissipate pulse energy and if the intake is poorly shaped in this respect, the engine will not respond to changes in intake length.

5 Carburettor

Some two-strokes have used constant velocity carburettors, but the vast majority use slide types; the choice being between those with round slides and those with flat slides. Both types can feature smooth-bore versions which, as the name implies, have a smooth venturi with the minimum of steps and interruptions. This type will flow fractionally more and will respond to pulse tuning.

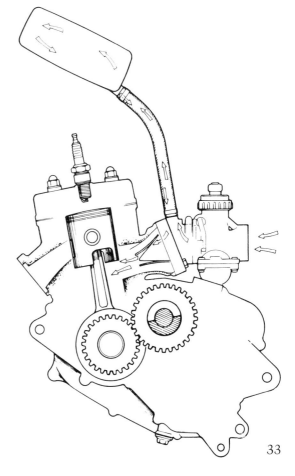

Fig. 12. Gas flow through a typical reed valve; suitable port sizing, or windows in the piston will leave the port open for a full 360 degrees, letting the reed valve dictate opening and closing periods

The size of the carburettor is a moot point. Some engines seem less sensitive to it than others and the general tendency is to use the biggest choke size that the engine will accept. As a very rough guide, 1.2 mm of bore size corresponds with each bhp that each cylinder produces, but there will always be exceptions.

There is something to be said for using the smallest bore size that you can get away with – it will give a wider spread of power because it will work down to a lower engine speed and still be able to atomize the fuel efficiently. It will also be easier to tune because the air will travel through it at higher speed and, at the top end, it will tend to richen up, which is safer than having the mixture go unpredictably lean.

For more details on carburettor selection and tuning, see Chapter 8.

6 Reed valves

There are several variations on reed valve control:

(a) A piston-controlled port, with a separate, reed-controlled by-pass connected directly to the crankcase. Even when the piston has closed the main port, gas can still flow if the pressure conditions allow it. Another advantage is that there is no reed block to obstruct the main passage although, of course, the port timing is subject to the same limitations as non-reed valve engines. The reed passage makes an awkward route and it all adds volume to the crankcase chamber, reducing its pumping efficiency.

(b) Disc valve intake, with extra port width before and after the main port, controlled by reeds. This enables the port to open early or close late, on

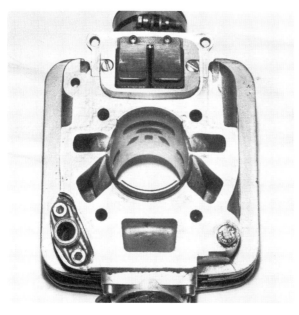

Fig. 13. Crankcase reed. On this KTM cylinder, the intake is piston-controlled with a reed-controlled by-pass directly into the crankcase

demand from the engine, while still having a smooth, unobstructed main passage. The reed chambers add very little to the crankcase volume. The disc timing can still be altered to suit new engine specifications.

(c) Piston-ported, reed valve controlling the full intake area. The early designs were no more than traditional piston-ported engines, with a reed valve inserted between the carburettor and the piston. The effect can be used in two ways. One, with porting etc., aimed towards low-speed torque and good flexibility, as in trials engines, the reeds can be used to boost these characteristics considerably. As an example, the 325 cc Suzuki trials engine, running without reed valves would give peak torque at about 5,000 rev/min. With suitable reed valving, it would give the same amount of torque but it could be made to peak at 2,000 rev/min.

When Yamaha developed their 125 motocross engines, they found that they could use the reed valve to increase power at both ends of the rev scale. With the reed valves, they could open the intake at 140 degrees BTDC, compared to 88 degrees BTDC on the piston ported engine. This gave an increase in power between 8,000 and 10,000 rev/min, and a gain of nearly 15 per cent on peak power. Through the midrange there was hardly any change, and below 5,000 rev/min the reed valves showed a further increase in power.

Yamaha's engineers went on to do a lot of work, optimising the dimensions of the reed block and choosing the best reed stiffness and lift. They came up with the now (almost) universal V-shaped reed block, with rectangular reed petals, and found that the reed valve area (projected along the line of air flow) should be in the region of 80 to 90 per cent of the carburettor bore area. Making the gas flow in what is effectively a tapering passage reduces the risk of turbulence. The fully developed reed valve engine gave about 30 per cent more power than the piston-ported machine.

(d) Reed valve controlled intake. Subsequent development of the reed valve in high-speed engines led to greater intake duration, until the intake port

Fig. 14. Types of reed block

Fig. 15. This Honda reed block has a plastic filler to take up dead space and direct the air flow in the required direction

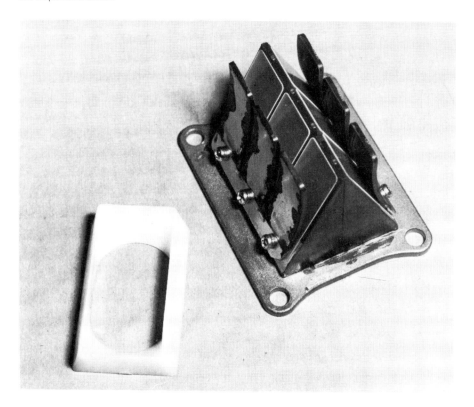

Fig. 16(a). Typical modifications to a standard reed block to increase window area. The area can be increased further by removing the central bar and using single petal reeds. For this to be effective it will be necessary to modify the rest of the intake tract to keep the optimum proportions (the reed block should be around 80% of the intake area)

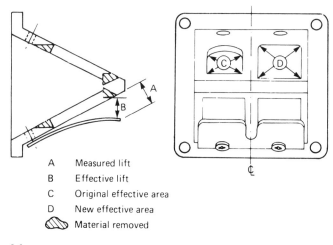

A   Measured lift
B   Effective lift
C   Original effective area
D   New effective area
 Material removed

Fig. 16(b). *Top:* By using a combination of two main jets and several different reed petals, on an RD250LC, the load and fuel flow could be varied considerably. The change in load (torque) is small at 9500 rev/min, and becomes greater as the speed increases

Bottom: By choosing certain combinations of reed and main jet the torque curve can be tailored, as these two extremes show. It would be possible to get several different torque curves between these two

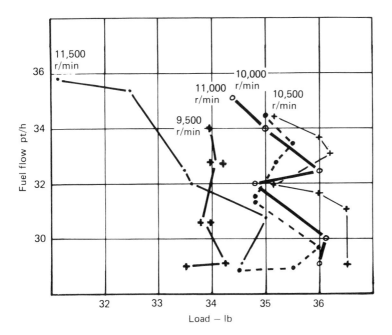

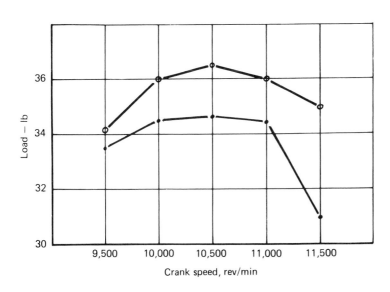

Fig. 17. Modified reed block

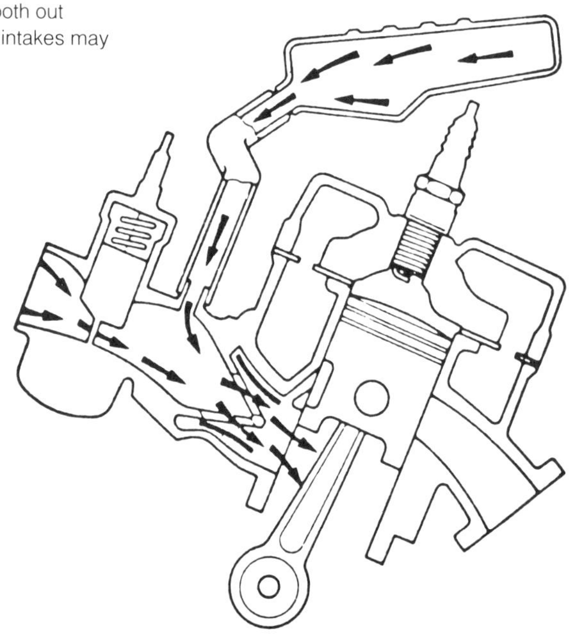

Fig. 18. An intake chamber can smooth out unwanted pulses; on a twin, the two intakes may be cross-connected

was left open for a full 360 degrees, the reed opening and closing as the crankcase pressure demanded. Variations on this theme include pistons which do actually close the port, but which have small windows in the skirt, so that at BDC the crankcase is open to the reed valve chamber.

Reed valve modifications

There are several things to consider; the intake time-area, the physical obstruction to air flow and compatible reed material. Modifications to the reed valve alone are unlikely to make any noticeable difference in peak power to a stock roadster – differences in the order of 2 per cent are fairly typical. However, it is often possible to shift peak revs slightly, by 200 to 500 rpm upwards or downwards and this can make a significant difference to the spread of power, especially on an engine whose power drops rapidly after peaking. In conjunction with other tuning steps this can be a useful way of tailoring the power characteristics. If the engine's air flow or its operating speed have been substantially changed, then there is every reason to suppose that the reed valve will also need to be altered to suit.

Intake time-area: where types (a) and (c) above are concerned, the time-area is mainly dictated by the piston and, to match increased air flow or speed, the port's duration can be extended by machining the floor of the port or the piston skirt, to make the port open earlier and close later. The area of the port can be extended by raising the roof of the port to match the piston at TDC. The port must not be raised so far that the piston rings can expand into it when the piston is at BDC. Type (c) can have its timing extended until it becomes type (d), usually by making windows in the piston skirt. Cutting away the whole skirt will shorten piston life considerably and where a large area has to be removed from the bottom of the skirt it may be necessary to machine the ring lands to prevent piston rock from damaging them (see Chapter 6).

Type (b)'s time-area is obviously influenced by the reed petal material, (same for the crankcase reed in type (a)) but it is easier to alter it by changing the opening or closing point of the disc valve, which is dealt with further on.

Obstruction to flow: the valve block and its petals obviously restrict flow in the port, and the necessary change in port section can easily induce turbulence in the gas flow. Simply making the chamber bigger and using a larger valve will not necessarily improve flow, it could create turbulence and actually reduce the total flow through the port.

The effective valve area is the window area with the petals on full lift, projected along the axis of the gas flow. As mentioned above, Yamaha found that the optimum valve area was 0.8 to 0.9 times the bore area of the carburettor. This corresponds to the conventional practice of tapering a port on the approach to an obstacle (such as the valve stem in a four stroke) because turbulence is much less likely in a nozzle than in a parallel or expanding passage. If the rest of the port is to be enlarged and a bigger carburettor used, then a larger reed block should be fitted, keeping to the same proportion.

Sometimes it is possible to find a larger valve, from a bigger engine or a competition variant and before this can be fitted, the valve chamber will have to be enlarged – with the risk of breaking through one of the walls if the alloy casting is not quite thick enough – to take the new valve. To avoid this it may only be necessary to take extra care not to machine too much material away; or to modify the new reed block to fit the chamber; or to build up behind the thin area with alloy weld *before* machining; or to fill in the damaged area with an epoxy plastic metal which can stand high temperatures.

An alternative is to increase the valve area of the existing block. This can be done by increasing the lift of the reed petals, by carefully bending the stops, although this is not always successful. On some engines it works, on others it causes a loss in power. For each reed material and thickness there is an optimum lift; overlifting can reduce power, possibly by making the reed flutter. It also seems that the curvature of the reed stop is important. If it does not follow the natural bending curve of the petal, or if the stop is twisted, side-to-side, then there can also be a loss of power.

Opening up the windows in the reed block will also increase the area and, at the same time, gives the opportunity to streamline the bars which sit across the gas flow. Leave an area of 1 to 2 mm for the edge of the petal to seat on and don't remove the black rubbery material on the seat – this cushions the petal, helping to prevent bounce and damage. Some tuners remove the central bar and use one-piece reed petals, which increases the area but may shorten the life of the petals.

Increasing the area and streamlining the shape of the valve will have maximum effect at speeds above peak torque, where the air flow would normally be tailing off. This might just prop up the power curve and prevent the engine from being too peaky.

Reed petals: there are two theories about the behaviour of reed petals, both to do with the stiffness of the material. The stiffer the material, the higher will the natural frequency of the reed be, for a given length of petal. One theory says that low frequency petals will be easy to lift but are more likely to go unstable and flutter, possibly not sealing very well at low speeds. The other theory is that high frequency petals will get into resonance with the engine at higher speeds, and will then take less energy to open and so will give a boost at this speed – the higher the frequency the higher the speed.

In tests which Yamaha published, (using an engine with 360 degree intake duration) they found that reeds with a very low frequency (62 Hz) would reach full lift at 3,000 rev/min and at speeds above this. By comparison, a stiffer reed (123 Hz) only opened about 70 per cent of full lift at 6,000 rev/min and didn't open as far as the low frequency reed even at peak speed. When engine tests were made, they found that the 62 Hz reeds gave a high-speed increase of 2 bhp and let the engine peak at 9500 rev/min which was 500 rev/min further up the scale, compared to some very stiff, 153 Hz reeds. Note that an engine speed of 9,180 rev/min is a frequency of 153 Hz, so the stiffer reeds actually peaked close to their resonant point.

At low speeds, the stiff reeds displayed the results of being able to close more rapidly and decisively: they gave about 2 bhp more than the 62 Hz reeds at speeds in the region of 4,000 rev/min. Yamaha also found that reeds with a frequency of 123 Hz gave almost as much power as the stiff reeds at low speed, and almost as much as the 62 Hz reeds at high speed. Through the midrange the difference between all of the reeds was negligible. They came to the conclusion that the best compromise was a reed whose frequency was about 80 per cent of the engine's frequency at peak power.

They were using stainless steel reeds, which are usually fitted as standard to roadster machines and to some racers, like the TZ Yamahas, although most moto-cross engines and practically all after-market reeds are the phenolic or resin laminate type. One difference is that the steel types are less likely to fail, although when they do, the results are usually more serious than when the other types break.

Tests which I have made on a variety of reeds have been less than conclusive. As the table shows, there is no real pattern in the small changes in power, or in the shifts of peak speed and it could well be that in a stock roadster the reed block isn't the restrictive component and that other items are dominating the power production to the extent that reed changes do not have noticeable effects.

Table 4.1 Reed valve stiffness v. power characteristics

Petal thickness mm	Petal stiffness gm/mm	Change in peak power bhp	Peak power speed rev/min	Speed range above 40 bhp rev/min
0.2 (steel)	28.3	0	8,500	1,430 (2)
0.34	30.7	+0.1	8,500	1,430
0.39	30.7	+0.7	8,700	1,570
0.50	30.2	−1.2	8,700	1,290
0.65	39.7	−0.2	8,700	1,640 (2)
0.2 (steel)	50.2	0	8,500	970
0.2 (steel)	50.2	+1.0	8,600	1,210 (2)
0.7	50.2	+1.3	9,000	1,430
0.7	50.2	+1.5	9,000	1,570 (1)
0.6	70.3	−0.4	8,500	1,210
0.39/0.58	104.0	−0.2	8,700	1,140
0.39/0.58	104.0	0	9,000	1,280 (1)

Note: (1) reed stops opened to 11.5 mm lift (standard 9 mm)
(2) streamlined reed block used.

If the modified blocks and the steel petals are regarded separately then a pattern does emerge, suggesting that the lower stiffness reeds give a broader spread of power at peak revs (shown in the last column). The stiffness figure relates to a small test rig in which the petal was clamped against a normal reed stop and loaded via a lever which pivoted on the same axis as the reed clamp, with a roller which pushed the petal down to the stop, the full deflection which the petal would reach in use.

Another series of tests (see Fig. 16(b)) showed similar changes in load and speed, plus small changes in fuelling which had previously been ignored. When the carburettor was adjusted to give the same fuel flow at a given speed, the engine tended to give the same load, no matter what type of reed was used.

The point here was that the reeds were having some effect on the carburation, presumably because the action of the reed would govern the shape of the intake pulse and this would alter the air speed at the carburettor.

The effect is there all through the rev range but around peak torque (where the air flow is at its greatest), the change in fuelling causes a change in torque of about 6%. At maximum speed, where the air flow is inclined to drop very rapidly, the effect is much greater – in the region of 13%. So although the gains in peak torque are likely to be small, there can be a worthwhile improvement in the width of the useful power band.

These tests also suggest that different reeds can be used to tailor the fuelling and the power characteristics. The machine used was a race tuned RD250LC which was fitted with seven types of reed valve, changing the main jets to optimize the fuelling in each case; this gave fourteen or fifteen different combinations of reed/jetting, some of which are shown in Fig. 16(b).

When the fuel flow is plotted against the dyno load (equivalent to torque), the resulting curves are very similar to the carburettor mixture loops obtained by making the mixture progressively weaker and measuring the load. In Fig. 16(b) the first three points (giving the highest fuel flow figures) were taken using 290 main jets; the remainder were taken with 270 main jets. The rest of the changes in fuel flow and load are due to the different reed valves. By choosing the best combination of reed valve and main jet, the torque curve can be optimized, which is also shown in Fig. 16(b). It would, in fact, be possible to produce several different shapes of torque curve in between the 'best' and the 'worst' cases shown. Normally it would be enough to get the 'best' torque curve – which is quite clear in this case. In other applications, the torque characteristics could be tailored to suit the needs of the engine, or of the circuit conditions. For example, if the maximum safe speed were 11,200 rev/min, there would be a good case for using the reed which made the high speed torque drop suddenly. On other engines there might be a choice between getting the highest peak torque or choosing the widest spread of torque (the same set-up happened to give both in these tests).

Finally, as well as influencing the width of the power band, different types of reed also had an effect on the way in which the engine came into the power band. Tuned two-strokes tend to stutter and four-stroke when they are part-throttled at speeds below the power band; a couple of the reeds seemed to suit the engine noticeably better than the others in this particular condition. They gave it a cleaner pick-up, making it smoother and easier to control. So far there does not seem to be any reliable way of predicting which type of reed will suit a given engine – but it is obviously worth experimenting with different types, particularly when the power band is narrow or when the throttle response is not as clean as it might be.

7 Intake tract

With the exception of the reed valve installation, the intake should be kept as straight and smooth as possible so that there is the minimum hindrance to gas flow and the pulses which travel along the port do not get broken up. If the port is smooth then these pulses can have some very helpful effects. They can also have harmful effects if they arrive at the wrong time and it will be necessary to experiment with intake lengths in order to make use of the good effects and eliminate the bad ones.

One problem is that the arrival of pulses depends upon the length of the tract (which is constant), yet the engine speed varies considerably. On occasions it is necessary to damp out stong pulses (which also make more

Fig. 19. A bridged intake port will reduce the risk of piston rock

noise) and it may be possible to make use of them to make the engine more flexible, or to come cleanly off idle when the throttle is opened.

A chamber located up close to the piston, or to the reed valve if one is fitted, will soak up arriving pressure waves like a surge tank. If the tank is the right size the same wave could be returned when there is something useful for it to do; alternatively it could be simply soaked up and dissipated. On a 180-degree twin, the connection could be made to the other intake, instead of a separate chamber. This sort of cross-connection obviously needs a lot of experimentation but it can make big differences to machines which have flat spots on low throttle positions, or have a rough idle problem.

The size of the intake tract close to the engine needs to be slightly less than the area of the carburettor, to avoid turbulence; where reed valves are fitted the transition from round to square section should be as gradual as possible. The reed valve chamber is part of the crankcase volume and any increase in its size will reduce the pumping efficiency of the crankcase. As it contains gas at crankcase pressure, it can be used to feed existing transfer ports, via cross-connections, or it can feed directly into the cylinder, via a 7th port, opening out above the piston (see Scavenge ports, further on).

Where a disc valve motor has reed valve compartments either side of the main port it may also be possible to connect these chambers with the scavenge ports, depending on the layout and the thickness of the casting.

Fig. 20. The bridge may also be horizontal

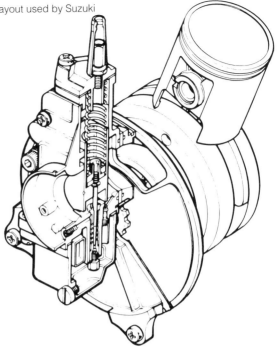

Fig. 21. Flat slide carb and disc valve layout used by Suzuki

Where the piston opens and closes the intake port, the window can be raised to match the piston skirt profile at TDC (or vice versa) but check to see that the rings don't extend into the port at BDC. If the ring pegs are on the back of the piston, this automatically limits the positions of any ports, because the ring ends obviously cannot be allowed to spring out into the port.

On some engines the rings do travel as far as the intake port window, in which case the width of the window will become critical to ring life. The current maximum is 70 to 80 per cent of the cylinder bore for race engines; less for roadsters, depending on how frequently you want to renew the rings. A central projection, down from the top edge of the port may support the rings – if your engine has one, don't remove it.

The gas flow from the port window has to pass the piston skirt. Cutting away the skirt will increase the time-area quite effectively and as pistons are cheaper than barrels, this is a good way to do any experiments. However, cutting a large portion away (where the piston once had windows, for example) will seriously shorten the life of the piston because the rear of the cylinder is usually the thrust surface and reducing its area will increase the pressure on the rest of the piston skirt. The lack of support may also let the piston rock and, instead of taking the thrust on its skirt it will push the ring lands against the cylinder wall, eventually smearing them and making the rings stick in their grooves (see Chapter 6).

Fig. 22. The component parts of Suzuki's disc valve intake

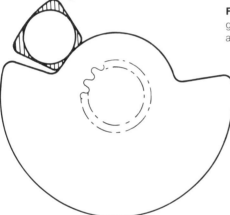

Fig. 23. A rectangular or wedge-shaped port will give more area than a circular port, without altering the disc valve's timing

The gas from the intake has to pass over the piston skirt, or through windows in the skirt and gas flow is more efficient over a knife-edge profile. Consequently a 45-degree chamfer on the skirt/window should improve flow into the crankcase.

As there are very few piston-ported machines made now, the intake time-area is not as significant as it is for the other ports. Reed valve engines can be developed to the point where the reeds control the timing; disc valve engines are easier to experiment with and to calculate the time-area; for piston-ported engines the best advice is to advance the timing gradually until blow-back at low speeds causes problems – alternatively consider the use of a reed valve. Where the piston does control the intake, the time-area can be increased in a number of ways. Making the port wider will increase the area without affecting the timing; raising the piston skirt will increase the duration, which has the same effect as lowering the floor of the port. Raising the top of the port will increase the area (assuming the new window shape is matched by the piston skirt) without affecting the timing. The program TA2A in the appendix can be used to calculate intake time-area. The relationship between intake and engine speed can be seen in Figs. 24 and 25; where time-area is concerned the intake needs rather more time-area than the exhaust, and up to twice as much as the scavenge ports, depending on how peaky the engine is allowed to be. If the intake is open for too long, low speed performance will suffer quite badly and the tendency to blow back through the carburettor will upset its mixing abilities, making the engine even more difficult to run.

This is yet another area where it is necessary to approach the optimum settings in small steps, at the same time trying to decide whether it is the intake or some other process which is restricting the engine's output.

If the restriction is elsewhere then changes to the intake will not show any improvement – but the bad side-effects will still appear. Engine tests, comparing load, speed and fuel flow, can often show where the restriction is:

1 A restriction upstream of the carburettor will cause the mixture to richen when the engine tries to flow more air. This will be most noticeable around peak torque (or where the engine is *trying* to produce peak torque).

2 A restriction at the carburettor will have roughly the same effect as closing the throttle slightly. The load will drop if the throttle is closed even slightly, which is not the case when the carburettor is too big, or when something else is restricting the engine. The air speed in the venturi will be at a maximum and the carburettor will be working at its most efficient, in terms of delivering fuel and of atomizing it in the airstream. The fuel flow might be slightly high for the size of main jet, especially at peak torque where the air flow is greatest. Air flow and torque will drop sharply after peaking; peak power and peak torque will probably be at the same speed. Midrange power should be at its best.

3 If the restriction is in the remainder of the intake tract, the symptoms can be much the same as for a too-small carburettor, except that the carburation will probably not be as good as there will be a greater tendency for the fuel to fall out of the air stream or for the carburettor to deliver 'wet' mixtures. A too-small tract will produce high gas velocities which will increase the bottom end of the power band. Conversely, a tract which is too large, or which is restricted by a poorly-shaped reed valve/chamber, will produce low gas velocities, with more tendency to blow back, upsetting the carburation and making the power band unnecessarily narrow.

4 Restricted crankcase or scavenge ports will cause the air flow to drop suddenly at whatever speed the porting becomes restrictive. Changes to the

intake will obviously produce no response from the engine; other symptoms depend on the nature of the restriction, whether the ports are physically too small or whether their timing is not matched to the other ports/engine speed range.

Fig. 24. There is a fairly clear relationship between intake duration and the speed at which peak power is produced. The difference between 'road' and 'race' specifications is due to the way in which other components are matched to the intake

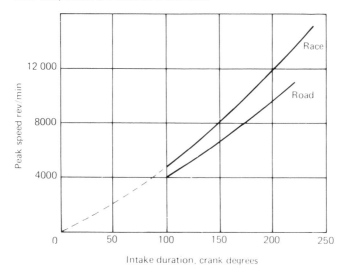

Fig. 25. The precise timing of the intake is not so clearly defined; the bold lines show the relationship between intake closing and peak speed. On piston-ported engines, this also defines the intake opening point; on disc valve engines the range of opening points is shown by the projected lines

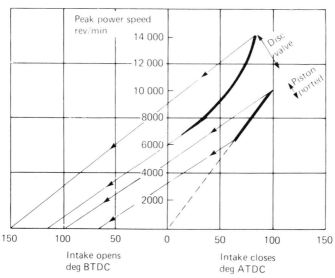

The final consideration for the intake port is that it usually aims the oil-laden gas stream at the big end bearing when it swings through TDC. Anything which happens to change this is likely to result in failed big-ends.

8 Disc valve

The advantages of a disc valve are the ability to use asymmetric intake timing, the immediate opening and closing of the port so that large time-area values are easily obtainable and the short, direct path into the engine. The mechanism generates extremely strong pulses and although the general wisdom is that the intake tract is too short to utilize them, they certainly have no difficulty in causing problems – mainly spray back out of the carburettor and carburation which is very sensitive to small changes.

The time-area can often be increased without altering the timing, simply by making the port shape match the edges of the disc and by making the port blend smoothly all the way into the crankcase. It should be angled to flow over the flywheel – which will usually be chamfered to streamline it as much as possible. The most efficient port shape, in terms of area, is a wedge, assuming there is enough material in the valve housing to provide this shape. Compared to a plain circular port opening and closing at the same time, a trapezoid port shape offers about 30 per cent more area. The corners should be generously radiused to prevent boundary layer drag impeding the flow.

The open timing of the disc is not usually critical. On high speed engines it will open just before the scavenge ports close – maybe this helps the final flow through the scavenge ports – at least it will prevent low pressure in the crankcase from stopping this final flow. The closing point is more important and is directly related to peak speed – the higher the speed, the later the valve closes. Some comparisons, along with peak speeds, are shown in Table 4.2.

Table 4.2 Intake timing

intake type	opens, deg BTDC	closes deg ATDC	peak speed
disc	115 to 118	37 to 55	8,000 to 9,500
disc	110	40	6,000 to 6,500
disc	145	65	10,000 to 11,000
disc	150	80	14,000
reed (360-degree port)	80	120	9,500
piston port	64	64	6,500
piston port	88	88	9,500

To mark up the disc for cutting, use a protractor or trace the outline of the disc on to graph paper; the centre can usually be found easily by projecting the lines of the opening and closing edges, or by drawing across opposite splines on the disc's mount. Add the required number of degrees (see the

Appendix for time-area calculations) and make a template from the drawing. The easiest way to cut the disc is to clamp it firmly between two pieces of wood and carefully saw through the wood and the disc with a hacksaw. Trim the sawn edges with a smooth file, taking care not to let the disc bend or distort.

The sensitivity of carburettors when used on disc valve engines seems to be related to their size, which appears to be critical to within about 2 mm. If the carb is too big, then it will probably be very difficult to jet correctly and may go from fully rich to fully lean in only a few jet sizes. Having watched disc valve engines run on a dynamometer it seems that the enormous fuel 'stand-off' may be responsible for this. On other types of machine it can be contained by making the intake longer, but this isn't practical when the carburettor sticks out from the side of the crankcase. It seems that a column of air vibrates back and forth through the carburettor, drawing off fuel with each pass. Perhaps there are some conditions when the engine collects all of this fuel and runs rich; if so, there may be other conditions when it loses the fuel and runs weak.

9 Crankcase

The crankcase acts simply as a pump, the volume displaced under the piston being forced up through the scavenge ports. The higher the pressure generated in the crankcase, the easier it is for the scavenge ports to do their job and the higher the compression ratio in the case, the greater its pumping efficiency. Also the speed at which peak delivery ratio occurs is inversely proportional to the square root of the crankcase volume. In other words, a small crankcase volume will work better at high speeds and if you want to raise the speeds, it would help to be able to make the crankcase volume smaller.

There isn't much that can be done to reduce the size of the crankcase, or to raise its compression (without obstructing the gas flow or preventing the cooling air stream from reaching the connecting rod bearings) but it is possible to avoid making the crankcase any larger. In this respect, reed valve chambers and wider scavenge port entries should be kept to the bare minimum.

Having said that, wide scavenge port entrances do improve the shape of the ports, because the reducing section of a nozzle is less conducive to turbulence. It is traditional to use the crankcase gasket as a template to make sure that the outlines of the scavenge ports in the crankcase blend smoothly with those in the barrel. In fact the evidence suggests that this makes no measurable difference (admittedly, it can do no harm, and I've yet to see a tuner who can resist doing it, just in case it does make a difference).

Many two-strokes rely on oil accumulating on the walls of the ports and running down into drillings to feed the main bearings.

This can sometimes be improved by counter-sinking the drillings, to provide a larger entry so they can accumulate more oil. It is also worth making sure that the drilling isn't obstructed by casting flash or by the edge of a gasket, particularly where a gasket is fitted under the housing of a disc valve.

10   Piston

The main functions of the piston are covered in detail in Chapter 6, but as it forms an important part of the valving and crankcase pumping systems, it is also included here.

The skirt, or windows formed in the skirt, control the timing of the intake and scavenge ports and consequently they can be used to alter the timing. The windows should be checked carefully to make sure that they align with the ports in the cylinder – when the engine is assembled, put register marks on the cylinder and piston crown, so that the piston and barrel can be accurately mated when the engine has been stripped.

Putting a 45-degree chamfer on the skirt and windows will improve gas flow over the edges of the piston and 'boost' ports, which allow gas to flow through the piston may also improve reliability by cooling the small end and the piston crown.

11   Scavenge ports

From the crankcase the gas is forced up into the cylinder, where its first job is to push out the remaining exhaust gas – but it must not mix with the burnt gas and it must not get lost into the exhaust system.

There are usually five or more ports, sometimes with interconnections between their entrances and to the intake (where reed valves are used). The directions in which they discharge gas are the critical factors.

When engines need good low-speed power, the scavenge ports usually aim towards the back of the cylinder, and upwards. Often the main ports will open before the auxiliary ports, or the top edges of the ports will taper down, so that the port opens progressively. This all gives good low-speed performance (and the sloping edges are kind to piston rings) with the minimum risk of fresh gas being lost to the exhaust. Giving the scavenge ports more duration and angling their exits up/back is often a useful way of filling in the midrange of a motor that is too peaky.

But, for high speed power, things have to happen a little more violently; there is less time for loss of the new gas and, within a narrow powerband, what does get lost can usually be retrieved by a resonant exhaust system.

So the main scavenge ports are angled to discharge across the top of the piston crown, and towards the centre of the cylinder. The auxiliaries are usually angled up – at anything up to 45 degrees – while the 7th port in the rear wall is angled steeply up. Originally any boost port would have been timed to open after the others, to give a final puff of fresh gas to wind up the

Fig. 26(a). Scavenge port modifications. *Left:* the dotted lines show how the port can be enlarged, with a window which is taller and wider than the original and with the gas stream aimed across the crown of the piston. The entry is made as wide as possible, smoothly decreasing to a minimum where the curvature of the port is tightest. *Centre:* the dotted lines show an optional modification where the gas flow is to be directed upwards. *Right:* proportions for a seventh (or 'boost') port, with the gas stream angled steeply upwards

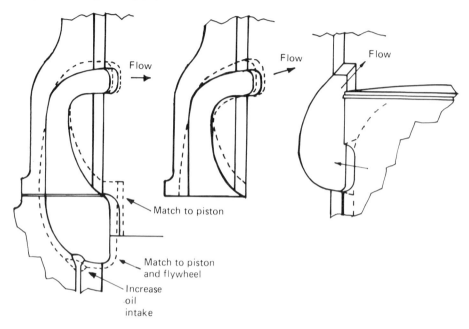

Fig. 26(b). The scavenge ports can also be modified to direct the flow of gas towards the centre of the cylinder (left) or to the rear of the cylinder (right). The direction of gas flow affects the power characteristics; gas directed across the top of the piston and towards the centre of the cylinder will tend to give better performance at high speed; gas flow aimed up into the cylinder, or at the back wall of the cylinder, will tend to give better low-speed performance

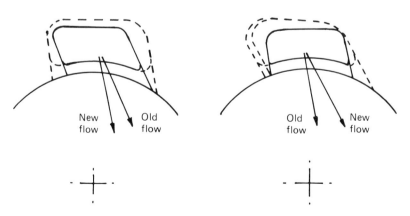

52

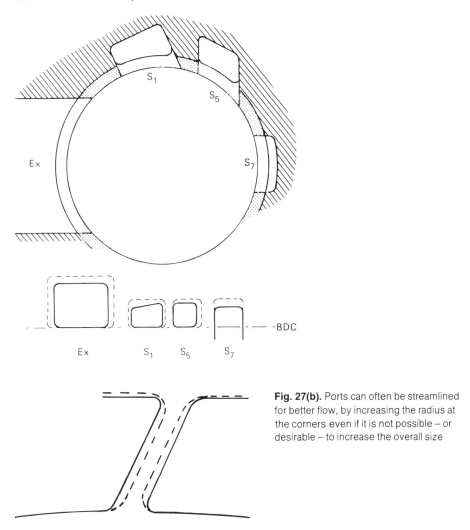

Fig. 27(a). Exhaust and scavenge port windows. The scavenge ports may be angled towards the rear cylinder wall, or may direct the gas streams to collide with one another. Typical port window modifications are shown by the dotted lines

Ex

S_1

S_5

S_7

-BDC

Ex S_1 S_5 S_7

Fig. 27(b). Ports can often be streamlined for better flow, by increasing the radius at the corners even if it is not possible – or desirable – to increase the overall size

scavenge process. The port can flow more than that, though, and, for high speed, it is common to open all the scavenge ports at the same time.

The only way to calculate scavenge timing for a given engine speed is to use pressure transducers in the cylinder and the cankcase (the manufacturers do this, or use computer simulations to predict the pressure changes). When the exhaust opens there is a sudden drop in cylinder pressure as the gas is free to escape (aided perhaps by abnormally low pressure in the exhaust port caused by pressure fluctuations in the exhaust system). The piston is still travelling away from the burnt gas, so the pressure drop can be quite severe. At the same time crankcase pressure is rising and any time after the case pressure exceeds the cylinder pressure, the scavenge ports can be opened.

So the first point is that the scavenge timing has to be phased with the exhaust blowdown, it cannot be considered as a single entity. If the exhaust time-area is increased then the blowdown will also be increased; the cylinder pressure will be lower when the scavenge ports open. Some increase in the pressure difference may be a good thing as it will make the scavenge gases emerge with more velocity. As all of the time intervals will be reduced when the speed is raised, an increase in gas velocity will (a) transfer more gas to the cylinder and (b) do a better job of scavenging the old gas. It will also increase the chances of new gas being mixed with the exhaust and of the new gas escaping into the exhaust system. For this reason the exit angles of the ports will become more critical but, in any case, there will be some dilution and losses at lower speeds so that power is bound to suffer in the lower two-thirds of the rev range.

Once the new gas is in the cylinder, some of its velocity can be destroyed by aiming the gas streams at one another; as they collide and slow each other down, the gas pressure will rise. Another way of slowing them down is to open the ports earlier (or to open one set of ports first) in order to start the scavenge flow before the pressure differential gets too high.

There is a third way which will pass roughly the same mass of gas but at a lower velocity and without changing the timing. This is to increase the scavenge port area, either by increasing the widths of the ports or by lowering the bottom edges until they align with the top piston ring in the BDC position (if the manufacturer hasn't already done this).

On racing engines the top ring is very close to the piston crown and gives precise control over the port opening and closing periods. Roadsters have a taller top ring land and this, while improving piston life, makes the exact opening position a bit vague. Some tuners file the piston crown away, forming channels aligned with the ports in order to increase the time-area and to give more precise timing. This will reduce piston life and the edges of the channels may become hot enough to cause pre-ignition; however it could be a useful way of determining the effect of more time-area without spoiling a cylinder barrel.

Some fine tuning of the ports may also improve the mass flow through them. In particular, anything which could cause turbulence should be removed, giving the ports a smooth, curved shape which tapers to a narrowest point where the curvature is greatest.

The scavenge gas needs to flow in a symmetrical loop, up the back wall of the cylinder and into the combustion space. If the ports are not exactly symmetrical, the flow will be lopsided and this will cause inefficient cylinder filling and poor combustion. There is a chance that swirl induced

Fig. 28. Modified RD Yamaha – the scavenge port entry is wide and the exit is angled across the top of the piston

Fig. 29. Multiplicity of ports in this Armstrong. The large pair are a bridged intake; the almost-square window is the 7th port; the other two are the secondary scavenge ports

Fig. 30. Kawasaki opened the reed chamber to the scavenge ports on some of their moto-cross engines

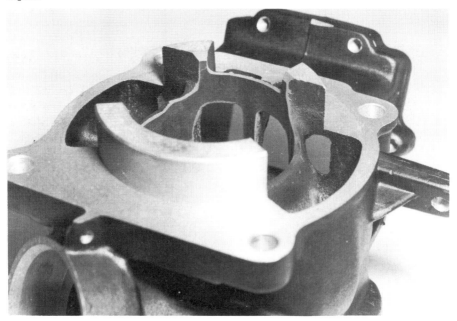

Fig. 31. The same Kawasaki as in Fig. 30 also has an extended intake port window, making a 7th port

by the scavenge ports *could* improve combustion but it would probably take a lot of experimental cylinders to find out if this could be put to any useful purpose.

It is possible to calculate the time-area values for the scavenge ports and for the exhaust (see Appendix), but, of course the figures on their own mean very little. The exhaust time-areas can be related to engine speed, but the

56

scavenge figures need to be related to the exhaust blowdown, to the engine speed range and to the number and direction of the scavenge ports. The same time-area values for ports pointing in a different set of directions would give totally different power characteristics.

The way to use the figures is as a comparator. To compare the engine with itself, the scavenge time-area can be compared to the exhaust, to get a 'blowdown ratio'. It can also be related to the time-area required at different speeds; that is, how much more time-area would be needed to get the same value at 9,000 as you had at 8,000? How much of this could be achieved without altering the blowdown relationship? Finally the time-area can be compared to another engine – to the sort of engine you want to turn yours into.

These figures should be treated as limits – they should be approached in stages, partly because this will give you more data to make more comparisons (it is hard to draw anything other than a straight-line graph if you only have two points to work with) and partly because there will be a need to match other components to the engine's changing state of tune.

There are a few practical difficulties when dealing with scavenge ports, particularly in smaller cylinder sizes. First, measuring the true width of the port is difficult when the port is set at an angle to the cylinder, as Fig. 32 shows. Putting a sheet of paper into the cylinder and letting the port window stain its outline will exaggerate the width of any port if you unroll the paper to take the dimension. This method will give the measurement along the perimeter of the cylinder, while the width of the port is the measurement along the chord – again as shown in Fig. 32.

Machining or filing the port shape is just as tricky; you need to make sure that you have tools to reach every part of the port – working on the accessible parts only is almost certain to spoil the shape, even if the window area is the desired size.

Measuring and marking the opening points is relatively easy, using a surface such as the top of the cylinder as a datum and converting the opening point into port height from this datum (see Appendix). A piston can be used as a guide to keep markings square to the bore and to ensure that all ports open at the same time.

The main ports will probably have a duration of 115 to 130 degrees and will exit at an angle of 90 degrees to the cylinder wall, aiming at a point just behind the middle of the piston crown. Aiming further back or higher up will tend to improve the midrange at the expense of peak power.

The auxiliary ports will be smaller – Yamaha mention a figure of 40 per cent of the size of the main ports – and will probably direct the mixture upwards, at about 40 degrees. There is no hard and fast rule about this; if it is possible to prevent gas dilution and loss of fresh gas, the ports can exit at any angle; the whole object of development work is to find a better combination of angles.

Additional ports in the rear of the cylinder may be the same sort of size as the auxiliary ports (in the case of disc valve engines) or smaller (7th port on reed valve engines). When these ports open at the same time as the others, they are usually angled steeply upwards, in order to keep the flow of gas in the loop format, and to prevent the stream of new gas being aimed at the open exhaust port which is directly opposite. When the boost port is opened later, it is often aimed directly at the exhaust, in order to provide an extra nudge in that direction and prevent any burnt gases avoiding the exhaust and continuing the loop back across the piston crown.

In general the scavenge ports need to be arranged to complement one another and to match the exhaust port timing. (Because the scavenge opening point dictates its closing point, this also has a bearing on crankcase induction and the intake timing.) Bearing in mind that at higher revs there is less time for the scavenge process to take place, the choices are:

1 *High stream velocity*
- high crankcase compression
- narrow port windows
- late timing (long blowdown period)
2 *Low(er) stream velocity*
- earlier timing; larger ports
- direction(s) of streams become critical.

The features which produce maximum peak power, with a narrow power band are:
- sudden port opening
- all ports opening together
- main stream directed across piston crown
- high gas velocity

The features which produce a wider spread of power, with a lower peak output are:

- progressive port opening
- staggered port opening
- gas streams directed rearward
- gas streams directed upwards

Of course, it is possible to combine some of these features to tailor the engine's power characteristics. Typically, the main ports would be arranged for peak power and the midrange then improved by carefully angling the auxiliary ports. All port windows should be chamfered or radiused, as described in Chapter 2.

12   Cylinder and cylinder head

These parts are covered in detail in Chapter 6, but as far as they affect gas flow, are included here.

The bore and stroke proportions have a significant effect on the available port area and the effectiveness of the scavenging loop, as mentioned in Chapter 3. While a long stroke offers more port area for a similar state of tune, the increased height and decreased width of the cylinder do not lend themselves to such an efficient scavenging process. As a modification, an increase in bore size gives more piston area which results in more power from the same gas pressure.

Cylinder heads which give high compression increase both the engine's pumping and heat-extraction efficiencies; also, the more violent the state of tune, the more susceptible the engine becomes to the effects of pressure pulses in the exhaust and intake. The disadvantage of this is that the power band becomes narrow and the engine becomes more difficult to silence.

A final feature used in cylinder heads is a squish band – of narrow clearance around the edges, see Chapter 6 – which causes turbulence before combustion and makes this process more efficient.

In some engines the forceful nature of the gas movement can be seen by the colouring of the spark plug electrodes. It is not uncommon for one half to be coloured light brown while the other half is dark, almost black in colour, suggesting that the plug is firing a very rapidly-moving stream of gas. This in turn suggests that the shape of the cylinder head, the positioning of the plug in it and the ignition timing would all have a bearing on power production. There was a tendency for manufacturers in the late '70s to make heads in which the main chamber was offset and the plug offset in it. Earlier designs had had wide squish bands and very deep combustion chambers – called 'top hat' designs because of their cross-section. More recently the movement has been back to symmetrical heads, with central plugs.

13 Exhaust port

This is the most dominant factor in determining the engine's power characteristics. Raising or lowering the exhaust time-area will have a significant effect on an engine even if no other change is made. If the rest of the engine is heavily restricted, the changes made by the exhaust will tend to be the undesirable ones – like making the power band narrow – instead of the desirable ones – like giving more power.

To achieve the full effect, everything else has to match the characteristics set by the exhaust so, after evaluating an engine and removing any obvious restrictions, the exhaust port is usually the best place to start any developmental work.

There is no point in opening the exhaust port while the gas is still at enough pressure to do useful work on the piston. In fact, if the tuning work is intended to raise the gas pressure then there is a strong case for leaving the exhaust timing, or even retarding it. If, however, the piston speed is increased then the expansion will eventually reach the point where the gas is not doing useful work on the piston and its remaining pressure would be better employed in helping the imminent scavenging process. At this point (from a pressure consideration) the port should be opened – and this implies that the opening point will be closely related to engine speed; higher revs will require more exhaust advance.

When the port opens it causes a sudden pulse of pressure to appear in the exhaust port and header pipe. This pulse contains energy which will be useful in scavenging the cylinder. The shape of the top edge of the port will

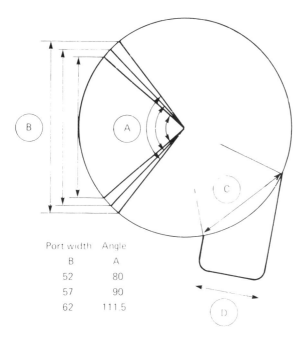

Fig. 32. Making the port wider leaves an ever-increasing arc in which the rings are unsupported. Above 90 degrees, the gains are not worth the penalties, as this diagram shows. The true width of a port (D) is not always the same as the width of the port window (C)

Port width	Angle
B	A
52	80
57	90
62	111.5

affect the production of this pulse – a port which opens gradually will produce a weaker pulse, spread over a longer period of time. A curved top edge will achieve this, as will a small hole bored just above the window into the port proper. The curved port edge also means that the piston ring will be brought gradually over the unsupporting region of the port window, where it will bulge out from the piston. And, more importantly, on the return journey it will be squeezed back into its groove in the same progressive way.

A straight, horizontal top edge will produce the strongest possible pulse, and the earlier it happens, the stronger the pulse will be. It will also inflict the most severe stress on the rings. As the pulse flies off into the intricacies of the exhaust system, the physical gas flow begins to emerge into the port. The rate of flow depends upon the pressure difference, so if the pulse effects from the previous cycle could leave a low pressure region in the port, this would boost gas flow considerably.

In this period, the gas is flowing out of the cylinder under its own pressure; the scavenge ports have not yet been opened. In this blowdown period the cylinder pressure has to be reduced to less than that in the crankcase, and at high speed there is very little time for this to take place. The port width becomes a factor in speeding up the blowdown, but there are several physical limitations to this.

First, piston rings do not last long enough when ports are too wide, because the less support they get, the more they bulge out into the port and the more severely they hit the top edge as the piston goes past. Even if this does not make them fail, the continual motion of springing out of the ring groove and being pushed back in causes fretting between piston and ring, and between the ring and its locating peg. This can cause wear, which will result in ring flutter, or it can loosen the peg which may decide to fall out, both being followed by fairly rapid failure.

Steady advances in piston ring materials and construction have allowed ever wider port openings (see piston rings, Chapter 6) and the practical limit for race engines is 70 to 80 per cent of the cylinder bore, for unbridged ports. Above this width ring life diminishes so quickly that there isn't any point. Even at this level, which leaves an arc of 90 degrees of the ring unsupported, ring life will not be long. In the recent past, engines had to use elliptically-shaped exhaust ports in order to achieve this width and keep their rings in one piece long enough to go racing. Piston ring technology now permits the greater area of a rectangular port to be used although, as with the top edge, a slight curvature will help the ring considerably, as will a generous radius where the top edge meets the vertical edge.

The port can be bridged, to support the ring in the middle and permit still greater widths – but this isn't a particularly satisfactory solution because the bridge runs very hot and is likely to expand against the piston skirt, causing a seizure. Usually bridges are ground to curve away from the piston (see Fig. 8) to avoid this problem, but in any case it is something which only the manufacturer can decide to put into a design.

Ring life isn't the only factor controlling port width. The second limit is the proximity of the scavenge ports. The closer the edges of the ports are, the more chance there is of scavenge gas getting lost into the exhaust port, particularly at low speeds when there is much more time for these things to happen.

Some engines extend the exhaust port over the tops of the neighbouring scavenge ports, making it a T-shape, or a wedge-shape, wider at the top than at the bottom. An alternative is to use a normal rectangular port, with holes bored at either side, level with the top edge, feeding directly into the exhaust port. The intention is to speed up the blowdown process, so that later exhaust timing can be used in an attempt to maintain a wide power band.

Exhaust blowdown must drop the cylinder pressure to below crankcase pressure before the scavenge ports can open, so, for optimum performance the exhaust and scavenge timing have to be carefully orchestrated.

If the scavenge ports are opened too early, then the new gas will not flow

Fig. 33. Cagiva bore two holes at either side of the exhaust port window

Fig. 34. The Kawasaki KX80 lets the exhaust port overlap the scavenge ports

out of them; there may be some dilution of the new gas; the crankcase may be pressurized by the burnt gases – which will increase the pressure under the piston and raise the engine's pumping losses. So there may be a few losses to the optimum power by opening the scavenge ports too soon – but only in the power band. At lower speeds the cylinder pressure will not be so high and so the scavenging process can start earlier. This is why an increase in scavenge duration can improve midrange power.

Both the scavenge and the exhaust timing are substantially symmetrical about TDC so the opening points also determine the closing points. (Some

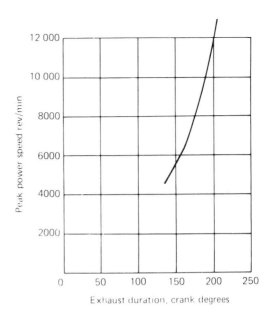

Fig. 35. Relationship between exhaust duration and the crank speed at which peak power is produced

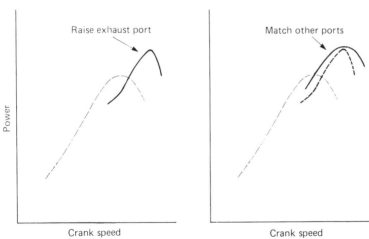

Fig. 36. The general effect of raising the roof of the exhaust port, on its own (*left*) and in conjunction with other modifications (*right*)

engines have offset pistons or offset bores, one effect of which is to make the porting very slightly asymmetrical about TDC.)

Ideally the scavenge should close when no more gas can flow from the crankcase. Opening the intake earlier may extend this period; the arrival of a very low pressure pulse in the exhaust port may also encourage more flow from the crankcase, with the likely risk that the new gas in the cylinder will also flow into the exhaust. There is then a period in which the scavenge

Fig. 37. Exhaust port windows employ a variety of shapes to maximise area and minimise ring wear

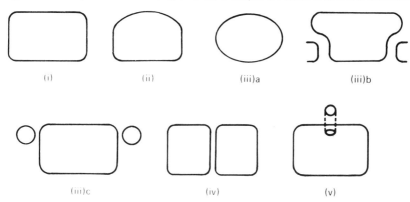

ports are closed, and the exhaust port is still open as the piston rises on the compression stroke.

During this period there is time to lose new gas or, if a pulse of high pressure arrives in the exhaust port, to prevent the loss of fresh gas and possibly even persuade some of the gas in the exhaust port to return to the cylinder. Obviously the generation and reflection of pressure pulses in the exhaust system have to be matched very closely to the closing time of the exhaust port.

Exhaust timing (or duration) is largely proportional to speed. If you take a random selection of two-strokes and plot a graph of exhaust durations versus peak speeds, it will produce a more-or-less straight line, going from 166 degrees/7,000 rev/min up to 200 degrees/12,000 rev/min.

Actually if you take enough of a sample, you'll wind up with three parallel lines, one above the line just mentioned, featuring bikes with very small cylinders/very narrow power bands/relatively small scavenge ports and a line below featuring more flexible, larger engines and bikes with big scavenge ports.

The exhaust time-area (see Appendix) is possibly the most significant calculation you'll need to make. The time-areas of other ports are modified by other factors, like the direction of flow from the scavenge ports, and the influence of reed valves on the intake. But the figure for the exhaust sets the whole specification for the bike. Like the exhaust timing it is closely related with peak engine speed and, put simply, it means that at higher speed there is less time, so the gas needs a larger hole. There is also less time for contamination or loss of fresh gas and the engine can get away with some very unlikely things. It doesn't get away with them at low speeds, though, and the price for a massive power boost at high revs is that the power band becomes very narrow.

The time-area calculation can be used to compare one engine speed with another (i.e. what needs to be done to get the same time-area further up the speed scale), or to compare one engine with another. The program gives both time-area and specific time-area, which is the time-area divided by the cylinder capacity, making it possible to compare one engine with another.

It seems that there may be some universal truths about exhaust time-area. Back in 1971, Yamaha's Hiroshi Naito published a paper in which he said they looked for 14 to 15×10^{-3} s-mm^2/cc in their GP racers. More than a decade later, a Kawasaki motocrosser, the KX125B2 gave 28 bhp at 10,500 to 10,700 rev/min; in this speed range its exhaust time-area was 14.7 to 15.5 $\times$ 10^{-3}s-mm^2/cc. The proportions between intake, scavenge and exhaust time-areas determine the power characteristics of the engine. A reasonable target to approach is 17–21 intake/16–18 exhaust/11–12 scavenge, approaching it in small steps and adjusting the proportions to give the required characteristics.

Fig. 38(a). Yamaha's power valve – a spool which raises or lowers the top edge of the exhaust port in relation to engine speed

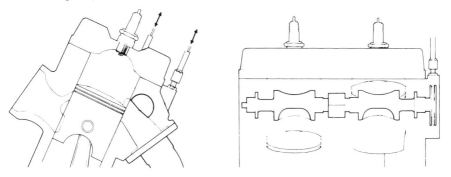

Fig. 38(b). Output curves for a Yamaha RD350F, with the power valve clamped in the down position (solid line) and then with it clamped in the up position (dotted line). When the valve is working normally, the operating mechanism cuts in between 6000 and 6500 rev/min and the resulting torque follows the solid line up to 6200 and from there it follows the dotted line

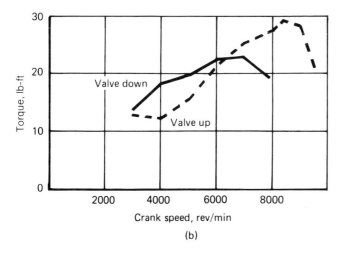

(b)

Fig. 39. Yamaha power valve

Figure 37 shows various exhaust port configurations; in each case the bottom edge should be lowered to match the piston height at BDC – preferably to match the height of the top ring, although this will encourage the top ring land to run hotter. All of the edges should be chamfered, as described in Chapter 2.

(i) The rectangular shape is the best for maximum area/high performance and is also the least kind to rings. The straight top edge generates the strongest pulses. The radiussed corner is the only concession to ring longevity and the radius should be at least 4 mm.

(ii) The progressive top edge is easier on piston rings where the piston is still at high velocity (unlike the lower edge) and doesn't produce such violent pulses. This may give a better spread of power and will be easier to silence.

(iii)a This goes the whole way, to get maximum port width with minimum effect on rings. Not as much time-area as (i) or (ii).

(iii)b Combining (i) and (iii)a where scavenge ports prevent full development of the port width, this arrangement speeds the blowdown period.

(iii)c Even where the port is at full width, faster blowdown can be obtained this way. The holes (which *should* appear as ellipses when drawn on a flat piece of paper) are bored straight into the exhaust port. Local overheating of the liner is possible.

(iv) A bridged port gives the ring support and allows a much greater port width, but the bridge tends to run hot and may cause piston seizure.

66

(v) A hole bored into the exhaust port from above the window will take some of the pressure and the violence out of the blowdown period. Used to detune engines.

The exhaust port itself needs to be smooth and free from steps or sudden section changes if the plan is to make use of pressure pulses. Steps will dissipate the energy in the pulses – and may be used deliberately in order to prevent an engine becoming peaky, to make it easier to silence or to make it less sensitive to exhaust system changes.

Because the exhaust port timing is so important, Yamaha produced a variable port, in which a spool rotates, with a cutaway section which fits up close to the top edge of the window. When the spool is rotated it has the same effect as raising or lowering the top edge. This is controlled by a small motor, triggered by a speed-sensing device which raises the valve for high speed power and lowers it for low-speed flexibility.

The torque curves in Fig. 38(b) show the precise effect, measured on the RD350F Yamaha. There is a substantial increase in torque above 6500 rev/min when the valve is raised, accompanied by a loss of torque below this speed – which is retrieved by lowering the valve.

Suzuki developed a mechanism which put the valve further away from the piston and had it open into a closed compartment which acts like a resonator chamber, see Fig. 40(a). The exhaust system only works beneficially over a narrow speed range; when the chamber is blanked off, the exhaust is allowed to boost high speed torque. When the chamber is opened, it prevents this effect – it also prevents the effects which whould cause a drop in low-speed torque, so when the chamber is opened at low speed, there is a useful increase in torque, as shown in Fig. 40(b).

Kawasaki combined both of these mechanisms, as shown in Fig. 40(c) and (d). In this system there are two auxiliary exhaust ports, which open before the main port. Inside the auxiliary ports there are valves, turned by a rack and pinion mechanism which is controlled by a simple, centrifugal governor. At high speed the mechanism opens the auxiliary ports, increasing the exhaust time-area and its duration. At low speed the ports are blanked off and one of the valves opens the main exhaust tract into a closed resonator compartment.

With the extra ports open, there is an increase in torque above 7000 rev/min, with them closed, there is an increase below 6000 rev/min.

The variable exhaust timing used by Yamaha and Kawasaki prove excellent models to demonstrate the connection between exhaust duration and the speed at which the engine makes peak torque. They also show how the time-area of the port affects the output of the engine. Table 4.3 shows time-area values for the KMX125, along with the torque produced when the auxiliary ports were fixed open and closed.

In both cases, peak torque occurs at a time-area value of 18.5×10^{-3}

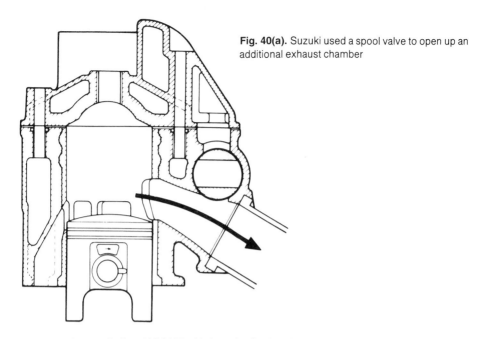

Fig. 40(a). Suzuki used a spool valve to open up an additional exhaust chamber

Fig. 40(b). Output of a Suzuki RG125 with the valve fixed so that the chamber was closed (dotted line) and then with the chamber open (solid line). The operating mechanism opens/closes the valve when the crank speed is between 8000 and 85000 rev/min

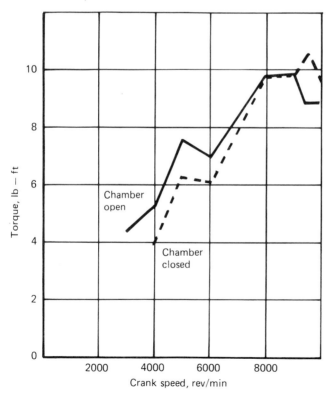

Fig. 40(c). Kawasaki developed a system with two auxiliary exhaust ports controlled by valves which were opened by a rack and pinion which was moved by a centrifugal governor when the engine speed reached the critical level. As well as altering the exhaust time-area, one valve also opened an exhaust resonator chamber to improve low-speed torque

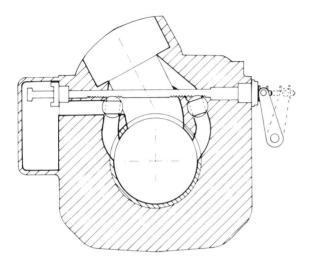

Fig. 40(d). Porting diagram used in the Kawasaki KMX125, showing how the main exhaust is augmented by the two auxiliary ports which not only add area but also open earlier

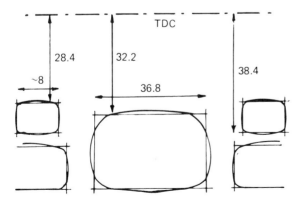

Fig. 40(e). The effect of the auxiliary ports on the KMX125 is shown quite clearly when the mechanism is fixed open (dotted line) and fixed closed (solid line)

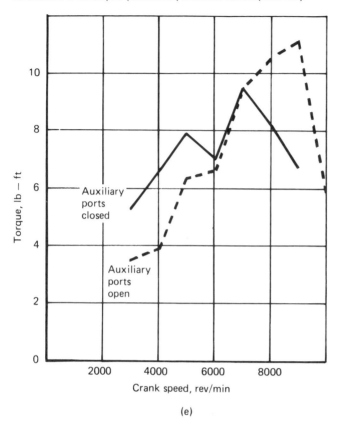

(e)

Table 4.3 Kawasaki KMX125: exhaust time-area and engine output

| Speed rev/min | Aux. ports closed | | Aux. ports open | |
	Torque lb-ft	Specific time-area s-sq mm/cc × 10⁻³	Torque lb-ft	Specific time-area s-sq mm/cc × 10⁻³
3000	5.3	43.2	3.5	52.2
4000	6.6	32.4	3.9	41.4
5000	7.9	26.0	6.3	33.1
6000	7.0	21.6	6.6	27.6
7000	9.5	18.5	9.5	23.7
8000	8.2	16.2	10.5	20.7
9000	6.7	14.4	11.1	18.4
10000	–	13.0	5.8	16.6

s-sq mm/cc; when the ports are closed, this happens at 7000 rev/min, when they are open, it happens at 9000. The duration of the main port is 176° which tallies with a peak speed between 7000 and 8000 rev/min. The auxiliary ports increase the duration to something in the region of 190°, shifting peak torque up to 9000 rev/min.

When the time-area values fall below 14×10^{-3} the load drops considerably – the exhaust is not open long enough for the engine to function properly. When the time-area is too great – with values much above 30×10^{-3} – the load also drops, because now there is ample time for fresh gas to be lost into the exhaust or for exhaust gas to find its way back into the cylinder. Between these extremes, the engine produces its natural power band, as the torque curves show.

The effect of the exhaust system can also be seen. Something, presumably the exhaust, is inhibiting the engine at 6000 rev/min; it works well at 4000–5000 and also in the region between 7000 and 8000. When the valves are closed, the resonator chamber is open, and this will tend to neutralize the effect of the exhaust, so the torque curve obtained when the valves were shut does not benefit from the power-boosting effects at 8000–9000 rev/ min. Without the exhaust system, the air flow and torque should correspond to the time-area. When this is 18.5×10^{-3}, the torque is 9.5 lb-ft with the auxiliary ports closed. When they are opened, the same time-area value was moved further up the speed range, to 9000 rev/min. If this had supported the same air flow, the torque here would also have been 9.5 lb-ft (which would have represented an increase in power because of the higher crank speed). Instead, the torque is boosted further, to 11.1 lb-ft, because the engine is now running in the region in which the exhaust system resonates.

The sudden drop in load above 9000 rev/min is probably engineered in order to prevent over-revving; there is still enough time-area to give over 8 lb-ft of torque at 10,000 rev/min. The fact that the engine is producing a lot less than this suggests that something else is causing quite a severe restriction – it could be in the intake, in the design of the exhaust system, in the ignition timing, or in some combination of all of these factors.

Table 4.4 Air flow development

A typical sequence of steps to increase the air flow through an engine, within a certain speed range, coupled with the work necessary to match each step (approximately at first, exactly at the end):

Main step	Secondary work needed
1 Choose exhaust time-area for required speed/load	1(a) make exhaust system to match new speed range, or use neutral exhaust such as short stub.
2 Match scavenge porting to blowdown period.	2(a) may need to repeat steps 1 to 4 in small increments to avoid a total mismatch.
3 Test for intake restriction	3(a) re-jet.
4 Remove restriction or increase intake time-area	4(a) re-jet, possibly change exhaust.
5 Raise compression ratio	5(a) reset ignition timing, use harder spark plug, uprated coils.
6 Fully develop exhaust system	6(a) re-jet; may need to modify scavenge timing to improve midrange
7 Fine tune ignition, WOT carburation and part-throttle carburation	7(a) may need to improve lubrication, clutch, etc.

Chapter 5
Exhaust system

Because the exhaust has such a pronounced effect on the output of two-strokes and because it is so easy to change, it has become a popular tuning device in its own right. A lot of formulae have been produced to conjure up dimensions but they all have to make assumptions about the speed of the gas flow and the speed of sound in the gas flow, so at best they can only offer a starting point from which the final exhaust has to be developed experimentally. It is just as easy to start with something which works on the original engine or which works on a similar engine and to progress from there.

In some cases there may be justification for using a neutral exhaust – a plain pipe which is too short to have any effect through the engine's speed range – just to be sure that exhaust effects are not obscuring the test results.

Tests with a plain pipe can illustrate (by default!) just what the exhaust system can achieve. If the length of pipe is varied it will, at some stage, produce an increase in torque, but only within a very narrow rev range. The carburettor may need retuning to get the best power and the fuelling will go very rich at peak power. That is, it will use a lot of fuel but there will be no other signs of richness. The reason is that the plain pipe can help the cylinder scavenging – it can help it so well that much of the fresh charge will be lost through the exhaust port and never see the combustion chamber.

A fully developed exhaust can do much more than this; it can scavenge the cylinder efficiently and somehow push back the gas that would have been lost.

To explain this, we must assume that the exhaust is able to produce both high and low pressures near the exhaust port, and to do it in time with the scavenging cycle.

The requirements are fairly simple:

during blowdown . . . low pressure at exhaust port
during scavenge . . . low pressure at exhaust port
before scavenge closes . . . lowest pressure at exhaust port
after scavenge closes . . . high pressure at exhaust port

As high pressure pulses are generated by the engine each time the exhaust opens, these would be reflected back in the form of low pressure pulses from an open pipe; if the original high pressure continued until it came to a solid wall, this would then reflect back a high pressure wave, some time after the low pressure wave.

The traditional expansion chamber does just this, starting with a plain pipe which exits into a diffuser – a pipe of increasing diameter. Some distance later the pipe contracts into a nozzle, with a narrow tailpipe; the throttling effect of the nozzle is supposed to slow the gas and raise its pressure and from this bottleneck, a high-pressure wave is reflected back along the pipe.

As well as getting the relative lengths right for the effect to happen at the appropriate speed, the chamber must also be capable of handling the physical gas flow. There are a couple of other considerations, too. The system must be able to fit the machine; it must be able to suffer burnt and liquid oil without getting clogged up; it must also be silenced. These requirements aren't so hard to meet – the system plus an efficient silencer can be put into a compact length and the pressure pulses don't seem to mind bends and corners as much as they mind steps or sudden changes in section. Silencing is relatively easy because the system needs a certain amount of back pressure for it to work, and it doesn't seem to matter whether this comes from the restriction of a silencer or from a nozzle and a narrow tailpipe. The amount of back pressure can be critical though, as it will increase the combustion temperature and too much will soon result in molten piston crowns.

As expansion chambers have been developed, irrespective of the original formulae to which they were constructed, certain salient facts have emerged. The guidelines formed by these are just as effective in forming a starting point as any other method. So, starting with the traditional expansion chamber, as shown in Fig. 41, there is a plain header pipe, a diffuser, a parallel section, a nozzle and a tailpipe. In the dimensions shown, L represents lengths measured from the piston, A represents diameter (assume for now that all the pipes are circular in section) and x represents the angle of taper of the cone.

Length

If the length of the pipe L_1 is progressively increased from a very short value, there is a region where the bhp goes up, then it falls again as the length is increased still further and more changes in length have virtually no effect. In the region where it increases the power, a shorter pipe corresponds to an increase in power at higher revs. To modify a system to work at higher speed, therefore, the front section needs to be shortened.

Despite this, most testers seem to agree that the more significant dimension is L_4 (to the point where the high pressure wave is reflected). Some insist that this point is actually part-way along the nozzle section; others move the tailpipe into the nozzle section to increase the effect.

However, changes in this overall length cause the strongest response from the engine. Once again, shorter corresponds to higher speed.

Various opinions have been expressed about the effect of the tailpipe but it seems that as long as the back pressure is high enough for the system to work and not so high that it melts pistons, the tailpipe is really irrelevant. Making it very long will obviously raise its restrictive value but when the whole system is working in unison with the engine, small changes to the tailpipe make no measurable difference. In most applications it is incorporated with the silencer anyway and it is this which produces the back pressure.

Volume

There seems to be a case for saying that larger/more powerful engines need a larger volume of chamber – presumably because the gas needs to be able to expand before it can be throttled in the final section. The front taper, x_1 has proved to be critical. It should be a shallow taper, of no more than 15 degrees included angle. Perhaps this is to avoid turbulence in the gas flow, or more likely to avoid the dissipation of the returning high-pressure wave, which obviously has to be channeled into the header pipe. If the taper is too abrupt here, or if the angle is too wide, then the whole system starts to behave like a simple, plain pipe.

Using a shallow taper and trying to get a large volume could mean that the chamber would become very long, so some manufacturers have produced chambers with two or even three diffuser tapers. As long as the changes aren't too abrupt, this appears to work and achieves the required volume in a much shorter length. A larger volume tends to move peak torque towards lower engine speeds.

The taper on the nozzle, x_2 has quite a different effect to the front taper. At its most extreme it would become a flat plate, i.e. x_2 would become 180 degrees – and this works when the length is right, but only in an extremely narrow rev range. Making it a proper taper reduces the effect on power but spreads it over a wider speed range, until if it just became a parallel pipe ($x_2 = 0$) then it would have no effect at all (but it would have it at *every* engine speed . . .).

So a steeper nozzle angle increases the effect but only over a shorter rev range. This may make the power drop sharply after peaking – which a shallow angle may prevent. The parallel section is a matter of convenience, once the diffuser has expanded to the required sort of volume, it merely needs spacing out to produce the right length to the nozzle. If A_2 is increased, giving a bigger volume, it tends to shift peak torque further down the speed scale; a smaller volume moves peak torque to higher revs. In tests these changes are often very small – either because they are almost insignificant (or even coincidental) or because some other feature is having a

Fig. 41. Main dimensions of, and variations on, the expansion chamber exhaust system. The diffuser angle X_1 must not be too large or the chamber will start to work like a plain pipe. System A uses two diffuser angles, permitting the required volume with a shorter length L_4. System B takes this a stage further and tapers the system all the way from the port – harder to make, but necessary on highly tuned motors. System C has a fully tapered diffuser plus a long, slow tapering nozzle for maximum power spread. System D introduces the first element of the silencer instead of the tailpipe; this could be extended into the nozzle section

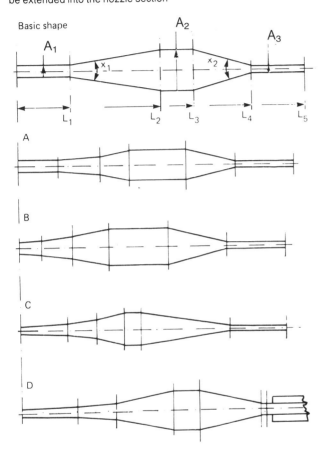

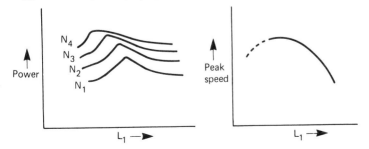

Fig. 42. The effects of changing the length of the front pipe: left, on power (N_1 = low engine speed, N_4 = high engine speed); right – on the speed at which peak power is produced. Some tests suggest that a very short (fully tapered) front section might reduce peak speed (dotted line)

Fig. 43. The effects of changing dimensions L_1 and L_4 on real engines (see Fig. 41). The scatter in the results is due to (i) different states of tune and (ii) the fact that changing the length can alter other properties, eg volume

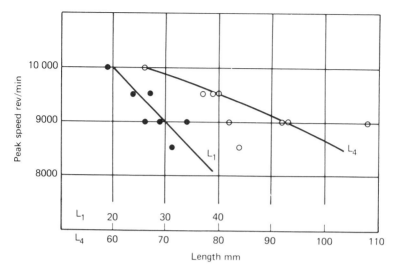

Fig 44. A variety of expansion chamber shapes and methods of fabrication ranging from (right) simple cones and parallel pipes welded together, through more elaborate cones and tapered front pipes, to a hydraulically formed system, which gives the smoothest shape, with the minimum of steps and sudden changes

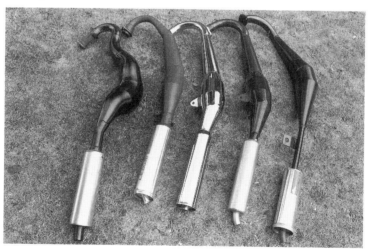

more dominant effect. Certainly some engines are more responsive to these changes than others.

The final development, which is the hardest to fabricate, is to replace the front pipe with a steadily expanding, shallow angle diffuser. This increases the effect on peak torque generated by the rest of the system and it allows the same results to be obtained from a slightly shorter system.

Fig. 45. Exhaust developed for a Formula 2 racer

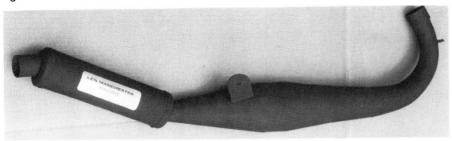

Fig. 46. Yamaha's 1985 500 cc GP racer; note fully tapered exhausts and complicated routing in order to fit required length on to the machine without losing ground clearance or increasing width

Fig. 47. Kawasaki motocross exhaust; full taper, large volume, short length, long tailpipe

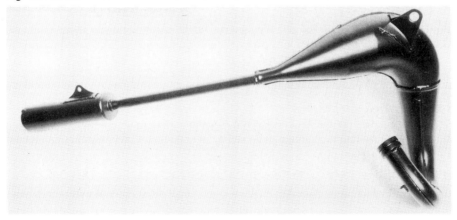

Once the basic length of a system is right, the chamber can be altered to tailor the power curve fairly effectively. In exchange for a slight loss in the peak value, the exhaust can give a sizeable boost to midrange power or (perhaps the most useful feature) it can carry on producing power well after the engine has peaked, instead of allowing the power to fall rapidly away. This can be useful because it makes the bike easier to gear and, for some circuits, easier to ride.

Silencing

For competition use, silencing applies only to the exhaust and maybe the piston in some air cooled engines, because so far the FIM noise tests are run on no load, at a speed based on the engine's stroke, so that all are measured at the same piston speed. Running the motor on no-load means that the throttle isn't open and there isn't any appreciable intake noise. Gas flow, and the energy it releases, is also minimal, so a simple, tailpipe silencer is more than adequate.

Street bikes however, are subject to a WOT drive-by test which means they need an efficient exhaust silencer, probably an intake silencer and also some means of reducing both mechanical noise and the reverberations from the casing of the expansion chamber. If these latter items work well, there will be less for the exhaust silencer to do to meet the sound level requirement. The EEC 1015 test highlights many of the facets of silencing.

Fig. 49. Absorption silencer, bottom, used in a street-legal exhaust with, top, a modified version used in a production racer

The test is made along a 20 m line, with the microphone of the noise meter placed 7.5 m to either side, level with the centre point of the line. Depending on its size, the machine must approach the line in second or third gear at a speed which puts it between 50 and 75 per cent of peak revs. At the line the throttle is opened wide, at the end of the line it is snapped shut.

Obviously the predominant noise is from the tailpipe, but significant amounts come from all of the other sources mentioned, plus the tyres, chain and transmission.

A liquid cooled engine is at a distinct advantage over an air cooled type – which will need ribs between its fins and rubber plugs inserted between the fins to prevent them ringing. The standard air box and its intake silencer will remove practically all intake noise, while an open carburettor at low revs on a two-stroke often makes as much noise as the exhaust. The chamber itself can give off a lot of noise and the Japanese manufacturers use pressed steel types, with welded seams down the sides, or double-skinned front sections to deaden the vibrations. Wrapping exhaust bandage around the front pipe will do the same thing and will prevent the same vibration from cracking the thin sheet exhaust material. A fairing will also remove a lot of this and the mechanical noise, although it tends to throw a lot of noise out of the back, which the microphone picks up as the bike goes away from it. Loose or dry chains will also add to the mechanical noise.

Finally noise is, in general, proportional to engine speed and the power curve can be tailored so that in the test conditions the engine's response is less active than it might otherwise have been.

All that is needed then, is an efficient exhaust silencer and, on a two-stroke, it should be possible to get down to the legal limit without making very much difference at all to the power production. The stock

Japanese systems are particularly effective in this respect, and if a basic starting point is needed, the bike's stock exhaust is as good a place as any.

Replacement silencers fall into two broad categories; the absorption type, which consists largely of a length of perforated tube wrapped in mineral wool or sound-absorbing material and stuffed inside a small can, and baffled silencers. The absorption type can have several problems. After it has been used a few times and the rock wool has got a bit oily it loses most of what effect it had. It may then start to raise the back pressure. The absorption type, as its name implies, has the same effect on all sound frequencies; it lops a bit off all of them. The trouble is that the exhaust noise is very rich in just a few frequencies and, even when they've had a bit lopped off, there's still a lot left.

The other type of silencer, of which there are several variations, depends on tuned lengths of tubes and small chambers whose volumes are also carefully adjusted so that each takes out a particular frequency. These are very effective – they can generate enough back pressure for the exhaust system to work as an expansion chamber, they can tolerate the oily contents of a two-stroke's exhaust and they are compact.

Fabricating exhaust systems

As a starting point, the stock system and any available replacement systems should be considered. A few quick experiments will show how different systems affect the power characteristics and it should be possible to choose something which is going in the same general direction as the rest of the engine development. Comparisons with the dimensions of competition chambers may also be worthwhile.

From this point onwards, a fair amount of cutting and welding is inevitable. It is possible to buy sections of pipe in various diameters and to get various 'bends' – a sheet metal shop may be able to supply these and some will be able to roll cones to your dimensions, which is a lot easier than trying to form your own. The traditional way of bending an existing exhaust is to cut a shallow V-notch, heat the bend and lean on it until the notch begins to close and then weld the edges of the notch together. To make a tighter turn it may be necessary to cut several notches (rather than one big one). Note that this shortens the pipe slightly each time and can leave ragged steps on the inner surface – which will not help the propagation of pressure pulses. For the same reason, any joints between sections of the chamber should be made to butt together as smoothly as possible on the inside and tacked in place before final welding.

One problem with making a pipe by cutting and welding is that any bends and tapers become very angular, with sharp corners projecting into the pipe. There is plenty of experimental evidence to suggest that if this kind of shape is severe enough then it will act like a baffle and will dissipate the pressure pulses in the same way that baffles inside silencers break up the sound pressure waves.

One of the advantages of dynamometer testing is that experimental systems do not need to fit the bike, they can be left sticking out at odd angles until the correct dimensions have been established, making fabrication that much easier. Then, having established the best dimensions, the prototype pipe is used as a model for one which will be shaped to fit the bike. Or, more often, it is given to a specialist with hydraulic forming equipment which can make smooth curves and tapers with no sudden changes or sharp projections. (See Figs. 46 and 66.) The system is cut out of flat steel sheet in shapes which will, when deformed, follow the required bends and curves of the finished system. This obviously needs a lot of skill and experience, which is why specialists are called for. The two sheets which will form the sides of the expansion chamber are laid one on top of the other and seam welded along their edges. A plain pipe is welded to the front to connect to a hydraulic line and a high pressure pump. The region where the pipe will eventually exit is welded up temporarily. Then the system is filled with water, pumped in at immense pressure – enough to make the flat sheet swell out like a balloon and, if the shapes are right, into the lengths and tapers of the prototype system but now curved to fit the lines of the bike. Complicated systems may have to be made up in several sections which are then welded together. The welded-up end piece and the front hydraulic adaptor are cut off and the cylinder flange and tailpipe are welded on: simple once you know how to devize the initial shapes.

Chapter 6
Piston, barrel and cylinder head

1   Piston

The piston is the most important single item in the engine, being the key part in both the pumping and heat extraction processes. As such it is the most highly stressed part and the one which is most likely to fail. Engine development was held up for a while when thermal loadings were greater than piston material could tolerate and the only way to make the piston strong enough was to make it so heavy that engine speed was then limited. Techniques in casting and machining alloys with very high silicon content solved this particular problem, because this type of alloy has low expansion and good anti-scuff properties combined with light weight.

The Japanese are particularly good at casting and their stock roadster pistons are often better than so-called competition replacement pistons.

(a) Piston clearance   This is one of the most critical factors to both engine performance and reliability. Too little clearance will result in friction, heat build-up, breakdown of the fairly scanty lubrication and seizure. Even if the lubrication keeps on top of the increased friction, power will be absorbed in large quantities – 10 to 15 per cent of the output is not uncommon.

Too much clearance will reduce the heat transfer from the piston to the barrel and may make the piston distort, risking seizure. It can also increase ring wear and allow piston rock which may damage the ring lands.

Finding the right compromise is not made any easier by the shape of the piston, which will be tapered, narrowing at the crown, and will be oval, not round. The reason for this is that the piston needs to match the shape of the bore when it is hot, and it isn't heated uniformly. The crown runs hotter than the skirt; the exhaust side is much hotter than the sides cooled by the intake and scavenge gases. To make it cylindrical at its working temperature, it has to take on an uneven shape when cold.

Measuring the clearance, when cold, can only be a formalised procedure, so it is essential that exactly the same procedure is followed each time. The official piston diameter is usually measured at right angles to the pin, at a point just below the pin. Japanese manufacturers usually specify a point, for example, 15 mm up from the base of the skirt. Similarly the bore is usually measured front to rear, 15 mm down from the top of the cylinder (check the exact positions with the factory manual). The official piston clearance is the piston size subtracted from the bore size; it will be obvious that this is not

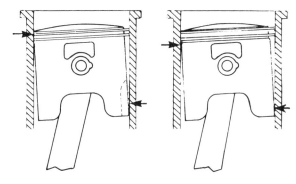

Fig. 50. Piston rock – caused by excess clearance or by removing part of the piston skirt. Left: if the thrust is taken on the ring lands, they will quickly fail. Right; machining the ring lands will cause the thrust to be taken on the skirt

Fig. 51. Large port windows in the piston skirt; on some racing engines the entire skirt is cut away, severely reducing the thrust area

an actual clearance because that part of the piston will never reach that part of the cylinder – or, if it does, the skirt clearance will have long ceased to be important.

The measurements need to be precise, so much so that the piston and barrel should be measured at the same time and place, to avoid any error due to temperature differences.

First the barrel should be checked for taper and ovality (see further on) and, if it isn't within limits, should be rebored or renewed. The clearance is then adjusted to the piston by honing (iron liners and some plated cylinders) or by selective assembly (plated cylinders). The manufacturer provides size-coded pistons to match cylinders.

If the piston has been modified so that the official measurement cannot be made, then a new procedure will have to be adopted. Before modifying the

Fig. 52. Indentation in the back skirt may relieve the point at which seizure is likely to occur

Fig. 53. Kawasaki changed the porting on the KX80. The E-1 (right) had a plain piston, the E-2 had a window communicating with a boost port and a stronger piston with a coated crown. Ironically the E-2 was peakier and seemed to suffer more piston seizures

piston, check its official clearance, and then measure the clearance using another part of the piston as close to the official position as possible, still at 90 degrees to the piston pin. If the official clearance was within limits (meaning that the piston and barrel are OK) then the new piston measurement subtracted from the bore measurement will become the new clearance, and from this point on, take all measurements in the same way on that engine.

The Japanese run very tight piston clearances on their road engines, partly because it keeps down noise and wear, partly because it gives better oil control and partly because they can. Unlike the European practice of relating piston clearance to bore size, and running air-cooled engines with a slightly greater clearance, the Japanese tend to relate the piston clearance to the output, as Table 6.1 shows.

Table 6.1 Piston clearances for various Yamaha machines

Model (year)		bore size mm	piston clearance mm	cooling system
DT250MX	(80)	70	0.035–0.040	air
DT175MX	(82)	66	0.040–0.045	air
RD400	(79)	64	0.035–0.040	air
RD350LC	(82)	64	0.050–0.055	liquid
RD350LC	(83)	64	0.060–0.065	liquid
RD500LC	(84)	56.4	0.060–0.065	liquid
RD125LC	(83)	56	0.050–0.055	liquid
DT125 MX	(82)	56	0.035–0.040	air
YZ125	(K)	56	0.070–0.075	liquid
RD250	(79)	54	0.035–0.040	air
RD250LC	(82)	54	0.050–0.055	liquid

Fig. 54. Suzuki coated the skirts of some of their pistons

In general an increase in speed and load will require a small increase in piston clearance, depending on the original clearance. The YZ125, with an output of 25 bhp and a rev limit of 12,000 rev/min has the highest specific output and speed of any of the models listed in Table 6.1 and its piston clearance indicates roughly what to expect.

Honing the barrel to give an extra 0.01 to 0.015 mm is a typical requirement, especially where the stock clearance is very tight. On models which already use something in excess of 0.05 mm, it might be enough to ensure that the clearance was on the upper tolerance and leave it at that. If subsequent tests showed power fade which couldn't be attributed to high combustion temperatures or an ignition problem, then the clearance could be opened up further. If a different make of piston is to be used, follow the manufacturer's instruction for setting the clearance. This will depend on the type of alloy used in the piston; those with a high silicon content (which includes most OEM) have a low coefficient of expansion and can therefore use a smaller cold clearance.

So far this has applied mainly to iron liners. Plated and fused bores give comparatively little choice and tend to run slightly tighter clearances. Kawasaki use an electro-fused finish on some of their bores and the piston clearance on their KX125 is 0.049 to 0.059 mm, on an engine with a comparable load and speed range to the Yamaha YZ125.

The smaller Kawasaki motocrosser, the KX80 E-1, which ran to 13000 rev/min had an even tighter clearance of 0.035 to 0.045 mm and this engine was famed for its reliability. In the following year, the E-2 model had a piston with a coated crown and a longer skirt, and the intake port was unbridged (tending to promote piston rock). The piston clearance was opened up to 0.040 to 0.060 mm and the engine was more prone to suffer piston seizure.

Where there is a problem with the ring lands smearing or being damaged and causing the rings to stick – which could be caused by excessive skirt clearance/cutaways allowing the piston to rock and take thrust loads on the ring lands, it may be necessary to machine the ring lands down – by the smallest amount possible, say 0.02 to 0.04 mm. Although the piston is oval, the clearance problems will only occur on the thrust axis, at 90 degrees to the piston pin, which is where the piston's diameter is greatest. Consequently the piston can be set up – gently! – in a lathe.

Other signs of piston overheating or distortion, leading to scuffing, may be cured by running a richer mixture and letting the latent heat of the fuel cool the piston. Tests run by Yamaha showed that, although a rich mixture gave less power initially, the power fade after several minutes running was halved. A liquid-cooled engine has much the same sort of advantage over an air-cooled one.

Increased lubrication, or the use of an oil with better anti-scuff properties may also help – see Chapter 9.

During the run-in process the engine should be stripped and high spots, indicated by signs of scoring or scuffing on the piston should be eased back with a smooth file. Persistent scuffing on the thrust face of the skirt can be cured sometimes by making a shallow depression with the tip of a countersinking drill bit.

(b) piston rings There are several types of ring used on high-output two-strokes, all developed to prevent ring flutter at high speed. The original, plain iron ring gave way to the Dykes L-section ring, which offered a number of advantages. First the working surface of the ring could be kept very close to the piston crown, while the groove was lower down. This gives good control over the port-opening but a short ring-land tends to run hot and

may fail, so being able to lower the ring groove is an advantage. Second, the ring is pressure-backed – that is, gas from the cylinder forces itself between the piston and the ring, which pushes the ring down on to the lower surface of the ring groove (preventing any tendency to flutter) and out on to the cylinder wall, making a better seal. Third, the ring is light (has less inertia) and, for its weight, it is very stiff and able to resist twisting or bending.

It was also quite expensive for production bikes and was generally superseded by the Keystone ring, a plain ring with its top face tapering by an angle of about 7 degrees. There was sometimes a smaller taper on the lower face. This is also pressure-backed to some extent and, as the piston ring bulged and contracted in and out of the ports, the groove clearance would continually change, preventing varnish and carbon deposits from building up in the groove. This self-cleaning action was quite a useful feature on high-mileage production engines.

For very high speed use, though, ring flutter was still a problem. It was eventually solved by using plain, thin rings, with a section which measured a thickness of less than 1 mm, whose lack of mass meant they could be used safely at high speed.

In the interests of reducing friction at high speeds, pistons with only one ring are often used. While this works well at high speed, it does allow blow-by at lower speeds, with the resultant burning and varnishing of the piston skirt which can only increase friction. It is really only a satisfactory answer in engines which are frequently overhauled and which are kept well within their dimensional tolerances.

Rings are located in their grooves by pegs, which are usually pushed into the piston. The safest type are those which sit in the ring groove and let the ends of the ring close over the top of the peg. The other, more commonly used, type consists of a brass pin forced into the piston, into the top edge of the ring groove. This type can loosen and come out, especially if the end of the ring frets against it as the ring moves into and out of the port windows. This type should be inspected carefully during each overhaul.

(c) piston stresses The acceleration reached by the piston at high revs is enormous (see Appendix for calculations of piston speed and acceleration) and the inertia forces generated by it are equally huge. The force is proportional to the square of the engine speed (so that doubling the speed actually quadruples the inertia loading) and it is this which puts a ceiling on the engine's safe operating speed. The same force is responsible for engine vibration, so an increase in speed in a single cylinder engine will cause a large increase in vibrational force – unless the engine has a balance shaft. If so, the balancer will counter the inertia loadings (unless a heavier piston is used, in which case the balance shaft would need a heavier counterweight) and the stress will be contained within the bearings and the castings between the two shafts. If the balance shaft is removed, this force will be

Fig. 55. Various ring sections used on two-strokes. The ring peg may be a pin pushed into the piston (*left*) or cast into the ring groove (*right*)

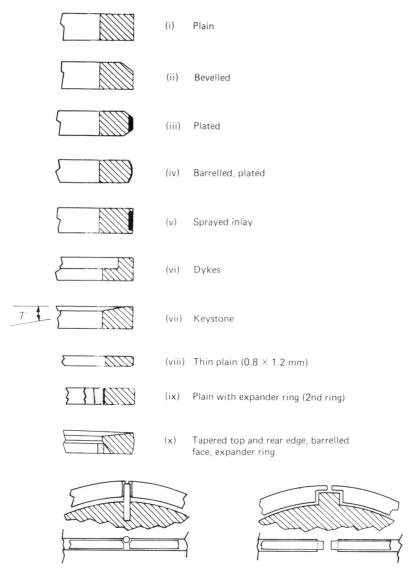

(i)   Plain

(ii)   Bevelled

(iii) Plated

(iv) Barrelled, plated

(v)   Sprayed inlay

(vi) Dykes

(vii) Keystone

(viii) Thin plain (0.8 × 1.2 mm)

(ix) Plain with expander ring (2nd ring)

(x)   Tapered top and rear edge, barrelled face, expander ring

transmitted through the crankcase castings to the engine mounts and the frame. Usually, one result of having a balancer is that the manufacturer can use light engine castings and frame fittings, so removing the balance shaft could easily over-stress these other components.

The connecting rod and its bearings also have to endure the stress produced at high speed and it may be necessary to use better quality parts – see Chapter 10.

(d) piston cooling The burning of oil under the piston crown is a guide to the temperature it reaches; once the appearance is charred the piston crown is receiving enough heat. The other dangers are localised heating under the spark plug caused by advanced ignition or a weak mixture; detonation, which will pepper the piston crown liberally; high combustion temperature, which will melt the piston crown, making it sag, or melt the top land in the region of the exhaust port; blow-by which will burn the skirt, caused by the ring not sealing or by excessive piston clearance.

Most of these overheating faults will cause very rapid failure. Where a fault is diagnosed, such as incorrect ignition timing, this can be rectified fairly simply, but as the engine is developed it may simply reach a point where the thermal loading is as much as it can stand.

There are ways of re-routing the excess heat so that the same level of power can be reached without damaging the piston. First, if detonation is involved, retarding the ignition and making the mixture richer will reduce the tendency, assuming that the motor is already being run on high octane fuel. Some two-stroke oils contain additives which either boost the octane rating of the fuel or stabilize it, so that mixing with the oil does not reduce the fuel's resistance to knock. With these oils (such as Castrol A747, Silkolene Pro-2 and Comp-2, Motul 300) the combustion is perceptibly smoother. The oil companies' competition departments are usually very helpful in supplying up-to-date information on both fuels and lubricants.

The way in which heat is conducted from the piston to the cylinder wall depends on the skirt clearance and the number of rings, as well as the lubrication. Excessive clearance will encourage the crown to run hotter and will produce blow-by which may burn the piston skirt. It may be possible to use a better grade of lubricant (see Chapter 9).

The two remaining alternatives are to cool the piston – by increasing the richness of the fuel/air mixture or by increasing gas flow under the piston crown (see Chapter 4) – and to cool the cylinder so that it will accept a greater heat transfer.

Make sure that there is no obstruction to the air flow around air-cooled engines. Ducting may help, and cooling the cylinder head may reduce any tendency for the fuel to knock. Liquid cooled engines usually leave little room for improvement, as long as none of the passageways are blocked. Some regulate the flow by restrictive holes in the head gasket.

2 Cylinder barrel

Two-stroke development has been forced to keep in step with the progress made in materials for the piston and barrel. The original iron barrels expanded at a different rate to the alloy pistons, which limited their usefulness. When iron liners were shrunk into cast alloy blocks things improved considerably, but there was still the problem of transferring heat from the iron to the alloy. No matter how good the fit, there was still a barrier at the interface of the two metals.

The Japanese perfected the process of bonding the alloy to the iron liner making a good thermal joint, which allowed a further leap forward in performance levels, aided by the use of high-silicon alloys for the piston. The ultimate move is to get rid of the iron altogether, and many attempts have been made at running plain alloy barrels, usually 'armoured' with a tough coating such as chrome, Nikasil, Galnikal and electro-fused layers which are exploded on to the surface by passing a heavy current through a wire held in the centre of the bore.

While there are advantages in the thermal performance of the engine in these methods, there are practical disadvantages, too. The cylinder cannot be bored to compensate for wear or damage due to piston seizure. Some can be honed, others have to depend on selective assembly to get the right piston clearance. Modifying port shapes is either difficult or impossible, as it will damage the protective coating. In some cases it can be done and slight damage made good, by replating the bore. Chrome plating is not so successful (and chromed rings must not be used). Nikasil is probably the best process. Mahle in Germany offer a reconditioning process and also supply coated rings for use with this surface treatment. The advantage is in light weight, better heat transfer and fewer problems with the relative expansion of cylinder and piston.

Wear While cylinders can be honed to give the correct piston clearance, they should also be checked for wear according to the maker's specifications. Usually this involves a sizing limit, plus limits for bore taper and out-of-round. The bore should be measured with an internal micrometer or a bore gauge (see Chapter 2) in six or eight positions: at 90 degrees to the piston pin axis, and along the pin axis; close to the top of the bore, in the centre and close to the bottom. If the total size is within limits, then the difference between any two measurements should not exceed 0.05 mm, or whatever the manufacturer has decided upon. Iron liners can be bored or honed to bring them back into spec but make sure that the operator can set the boring bar square to the line of the cylinder, and doesn't just follow the existing hole. Ask him to stop about 0.03 to 0.04 mm short of the finished size, so that it can be honed out to the final dimension, giving a good finish to the bore at the same time as setting the precise piston clearance. The best finish is a cross hatch at 60° to the cylinder axis, with a broad plateau area.

There are two basic approaches. One is to leave a small plateau area, which creates high local pressure but which has good oil retention. This needs careful running-in and is probably the best way to achieve maximum ring/bore life. Most tuned two-strokes have a short ring life anyway and it is common practice to hone the bore to give a broader plateau area, which can be run in quickly, using the engine relatively hard in short bursts. Any high spots which pick up or smear the alloy on the piston skirt should be eased down with a smooth file.

3 Cylinder head

The raw requirements for a combustion chamber are a compact volume of gas, equi-distant from the spark plug, promoting enough turbulence to help combustion but not so much that heat is lost to the metal surfaces.

A sphere with the spark plug tip at its centre would fulfil most of the requirements, especially as a sphere has the minimum surface area for the volume contained and any heat loss has to be proportional to surface area.

Bearing this in mind, it isn't surprising to find that one of the first developments away from the simple part-spherical head with flat-topped piston, was a head with a deep, central chamber. This early type had a broad squish band, leading into a deeply recessed volume of whatever was

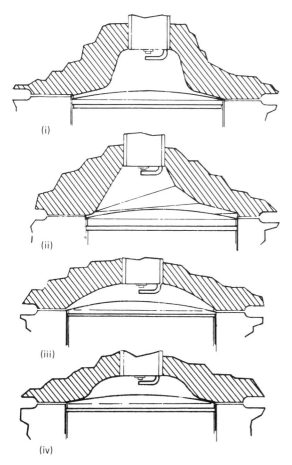

(i)

(ii)

(iii)

(iv)

Fig. 56. Types of cylinder head. (i) top hat section, with broad squish band and deep chamber. (ii) offset chamber, to use swirl created by scavenge streams. (iii) spherical. (iv) squish – carefully adjusted squish band, close-to-spherical chamber for minimum surface area

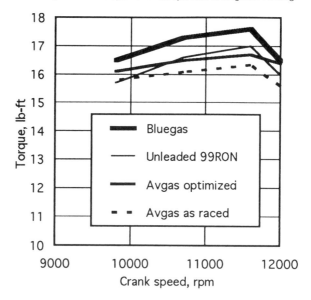

Fig. 57. The effects of optimizing a Honda RS125 on different fuels. To get the best torque curve it was necessary to change the cylinder head, first raising the compression to take advantage of higher octane ratings but also to alter the profile, making small changes in the width of the squish band. It was then necessary to find the optimum main jet size and ignition timing

necessary to keep the compression ratio within the bounds of detonation. It was called a 'top hat' head because of its extreme cross-sectional shape.

Later types tried offset chambers but experimenters eventually reverted to the symmetrical head and central plug, although with a narrower squish band, less deep chamber and gentle curves. And less surface area. Piston design has remained much the same, keeping the slightly domed, catenary curve of the crown.

What experimenters did discover was that the clearance around the edge of the head was critical to performance. In almost any two-stroke it is possible to get the compression so high that knock is a major problem, so the theoretical advantages of better thermal efficiency, etc., cannot be exploited anyway. But there were improvements to be had in combustion – making it burn faster so more heat goes into expansion and less into raising the temperature of the engine.

When the piston was run so that it nearly touched the edges of the cylinder head, it created 'squish turbulence' at the edges – forcing the gas in towards the central spark plug, at a time when the flame front was expanding out towards the edges. It created the necessary last-minute turbulence to speed up combustion, and it got rid of the 'end gases' – the gas which would have stayed around the edges, which the flame would have reached long after lighting the rest of the gas and which probably wouldn't

Fig. 58. Plain combustion chamber, offset plug (Kawasaki)

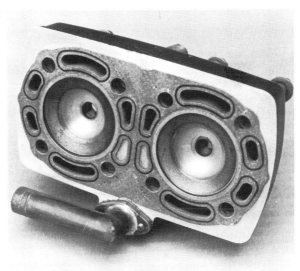

Fig. 59. Broad squish band – (race tuned Yamaha)

burn too well anyway, because there was so much relatively cool metal around it.

All that was left to be found was the optimum dimensions for this head shape. The central chamber can be designed on the basis of a sphere, to keep the surface area to a minimum, and it obviously has to contain sufficient volume to make a sensible compression ratio (see Appendix). Also the curvature of the surfaces needs to have as large a radius as possible, so that

93

Fig. 60. Plain chamber, central plug (Honda)

Fig. 61. Squish plus spherical chamber (Suzuki)

the squish effect can promote as much movement as is needed. That just leaves the squish area. The width of the band will be dictated more or less by the chamber size and its curvature. On a 56 mm bore it will probably be at least 8 mm across, and surprisingly, that accounts for just over 50 per cent of the piston area.

The critical part is the squish clearance, the gap between the head and the piston at TDC. The head must be profiled to be at least parallel with the piston and perhaps taper slightly away from it towards the centre. The minimum clearance to be effective is about 0.065 mm; some tuners run the engine so that there is just evidence of the piston touching the head at maximum speed. Squish clearance is usually measured by trapping soft solder between the piston and the head while the engine is turned through TDC. The thickness of solder is then measured.

94

A squish band can be increased by machining the head gasket surface and reprofiling the combustion surface in order to get the right clearance and adjust the compression ratio. Alternatively, alloy weld can be used to fill in the head, and the whole thing profiled, using a mill, or a lathe if the head can be held easily enough. The gasket face will have to be machined flat afterwards.

It is hard to see how the squish head can be developed any further on its own. One possible improvement might be to lower the whole profile and to use a piston with a concave crown, instead of the usual convex shape. It would help to be able to project the plug electrodes further into the chamber, if materials which can stand the conditions become available.

Raising the compression ratio to the limit of fuel knock is a standard tuning procedure. On a two-stroke it is a matter of juggling with ignition timing and the fuelling in order to get the best compromise between power, the shape of the power curve and no knock. The difference between tight, twisty circuits and those with long, full bore straights might make it worth while having different set-ups for each.

While raising the compression is supposed to be a matter of increased efficiency and is therefore applicable all through the speed range, it does, in fact, make peaky motors even more peaky. This is because the difference between being in the power band and being out of it is so great in terms of gas flow, that out of the band there's hardly any gas to compress. As far as the cylinder head is concerned, it's like running on part throttle. Raising the compression will improve the power here, but 10 per cent of nothing is still nothing. Meanwhile in the power band there is much more to work on, and the power will go up still further. The result is that at low revs there's no perceptible difference, but as soon as the motor hits the power band, there's an extra ten per cent, and the rider feels that as ten per cent more peakiness.

As a two-stroke's trapping efficiency varies considerably through its speed range, the head and the ignition timing have to be optimized around peak torque, where the charge density is highest. This means that combustion will be worse than necessary at other speeds. There is a head designed to overcome this problem, called the Polini Powerhead, in which the combustion chamber is cast as a unit which can slide up and down in a cylindrical bore. Its limit of travel is adjustable and controlled by a screw. The region behind the combustion chamber carries hydraulic fluid and is connected to a small, remote chamber in which a diaphragm separates the fluid from nitrogen gas under very high pressure.

The pressure of the gas and the position of the combustion chamber can be adjusted to give a high compression ratio and maximum thermal efficiency at speeds where the engine's volumetric efficiency is not very high, thus increasing the torque at these speeds. Normally this would cause detonation at speeds where the volumetric efficiency is high, but the increase in compression pressure pushes the combustion chamber away, against the

Fig. 62. Variable compression cylinder head. The chamber above the working part of the head is filled with hydraulic fluid under pressure from nitrogen gas in the remote chamber. If cylinder pressure is enough to overcome this, then it pushes the head away from the piston, reducing the compression pressure.

1. Cylinder head. 2. Sealing ring. 3. Hydraulic fluid. 4. Restrictor jet in fluid line to provide a measure of damping. 5. Earth lead for spark plug. 6. Diaphragm between fluid and nitrogen. 7. Pressure gauge. 8. Valve

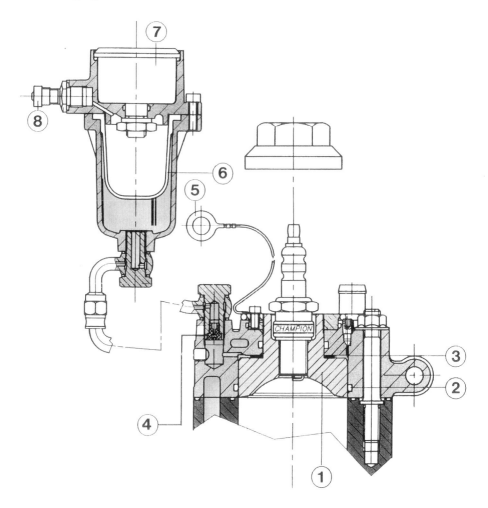

gas/hydraulic fluid pressure, reducing compression below the point at which detonation occurs.

Head gaskets vary. Some single cylinder motors don't need them, as long as the head and barrel are lapped together, or lapped on flat plates. The thin, strip metal types can be used – these are useful for packing up the head to adjust the compression height. The more exotic types include gas-filled

96

rings, and V-section gaskets which use gas pressure to force them against the surfaces and make an even tighter seal. Liquid-cooled two-strokes often need a composite gasket, with a central metallic ring to seal the head and gasket material to seal the water jacket. Make sure that the water passages through the gasket are not obstructed and that the head is torqued down evenly to give a good seal.

When the cylinder and head are clamped by their bolts or studs, they (and any gaskets) are compressed, while the studs are stretched. When the engine is run up to its operating temperature, thermal expansion increases the tensile force in the studs and the compressive force in the head. The force of combustion increases the studs' tensile loading further. The result is that the studs stretch like springs and this reduces the clamping force at the head gasket, possibly below the level at which it will seal. A fairly thick, soft gasket material could be elastic enough to compensate for these fluctuations but the cold clamping tension must be increased – possibly to the point where the gasket is permanently deformed and loses its elasticity. Combustion forces still stretch the studs, but now the gasket cannot compensate and either leaks or suffers a fatigue failure. In such a design it is important not to overtighten the studs, to retighten them when the engine has first reached operating temperature (in case of plastic deformation at the gasket) and not to reuse the gasket. Another approach is to use a thin, hard gasket (usually steel) and to choose stud sizes so that the required clamping force takes the stud close to the elastic limit of the material, to reduce fluctuations in the clamping force. The advantage is that the studs do not need to be retightened after the engine has been run for the first time and the gasket material is not likely to be taken beyond its elastic limit, causing a future failure.

Chapter 7
Ignition

There are three basic types of ignition system available for two-strokes: contact breaker (with either a battery and coil or a self-generating magneto), breakerless inductive discharge, and breakerless capacitive discharge, the latter being the most common. There are several ways of triggering the spark: contact breakers, Hall effect triggers, light-sensitive triggers, magnetic triggers and pulser coils but the essential difference is in the way the high voltage is produced.

Contact breaker types, whether incorporated in a magneto or driven by a battery, are limited by the switching speed of the contact breaker (this and its cam have a fairly low speed limit for delivering reliable sparks). At some point there will not be enough time for the current to rise to a sufficient level to produce enough kV to fire the plug; the voltage at the plug will tend to decrease just when it needs to increase. Having said that, plenty of contact breaker engines run to speeds of well over 10,000 rev/min. It is sometimes possible to extend high speed operation by using a cam modified to give a longer closed dwell, simply to increase the rise-time.

The other two, electronically controlled, types don't have this limitation. Where they vary is in the intensity of the spark. Both will deliver high voltages; basically the voltage at the coil will rise until it breaks down the resistance at the plug (or anywhere else it can find). But there is only a limited amount of energy, so the higher the voltage the less the current, or in the case of capacitor discharge, the shorter the time.

Now depending on the operating conditions, the plug needs a certain minimum voltage to fire it. The spark also needs a certain amount of time before it can light the gas, so the idea of spark duration can be important. Most stock ignition systems cope perfectly well with the stock engine, but often they don't have much in reserve and if the demands are increased, then they can't deliver. Sometimes it results in a noticeable misfire; usually it creeps in undetectably, the only symptom being a lack of power. As it happens at very high rpm, this often looks like the engine has peaked, and that's that. Sometimes, when the power curve is plotted, the power peak looks very ragged, which is exactly what it is.

When the ignition system reaches its limit, it misses out maybe one spark in every few hundred. Increase the demand and it misses two sparks in every few hundred – or the sparks don't have the energy to fire the gas. It is a gradual process which doesn't appear as a misfire, but soon it is taking a big

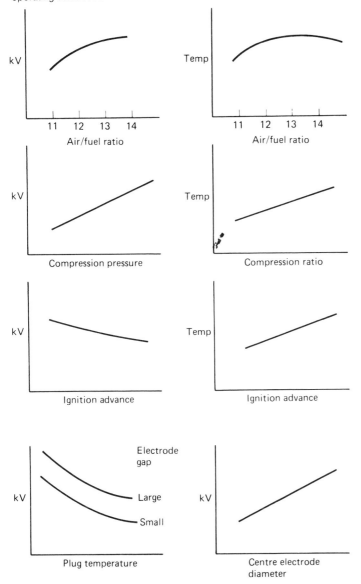

Fig. 63. Spark plug requirements; the variations in spark voltage and plug temperature with other operating conditions

piece of the power output. It also prevents the engine from responding to other changes.

When a more powerful system, or sometimes simply a better coil, is fitted, the normal power characteristics return. The choice is to try a better coil (a high performance coil, or one of the car competition coils), use an aftermarket ignition system, or fit one of the racing magnetos. One other problem with stock ignition systems is that they often retard the timing quite dramatically at a speed just above peak power, or else they have a rev limiter built in. If this is the case and you want to raise the engine speed, there is no option but to fit a different system.

Some systems have an advance curve built in – which may not be any good if the engine characteristics are changed drastically enough. Here the choice is to set the timing for peak power, and live with whatever it gives at other speeds, or to use a fixed-timing system.

The optimum timing is to fire the plug far enough before top dead centre so that the mixture can burn and maximum pressure build up on the piston as it gets maximum leverage on the crank's downstroke. Obviously if the timing is too late, the power delivery will suffer and if it is too early then pressure will build up on top of the piston before it gets to TDC and a fair amount of noise will arise from the cylinder head.

Unfortunately it isn't quite as simple as that because there is a lot of latitude between these two extremes. If you could run the engine at constant speed and slowly advance the ignition timing, you would see the power slowly creep up as well. You would also see the combustion temperature creep up until there was a real danger of knock or piston damage. The trick is to forsake this small power increment and to retard the ignition until just before the point at which the power really takes a dive.

The spark plug has some bearing on this because the state of its electrodes and the rate at which it can lose heat affect both its voltage requirements and the temperature reached at its electrodes. This alone may be enough to cause pre-ignition.

According to NGK, there are many factors which govern the plug's voltage requirements, most of which will be altered to some extent when the engine is modified. The voltage needs to be higher when the fuel mixture is chemically correct but decreases slightly when the mixture is richened; the plug temperature also decreases in this condition.

The voltage rises in direct proportion to the pressure inside the cylinder and so does the plug's temperature. The voltage requirement will reach a maximum at peak torque. The necessary increase in voltage with pressure is the reason that the ignition system cannot be tested simply by resting the plug on the cylinder head, spinning the motor and watching for sparks. All that tells you is that there is continuity in the circuit and that the plug isn't fouled or shorted. At atmospheric pressure the spark should be able to jump a gap of about 6 mm measured on a spark tester, depending on the ignition system. The only way to test a plug is in a pressure tester which will indicate the pressure at which misfiring occurs.

Advancing the ignition will actually lower the voltage requirement at the plug, for the simple reason that the compression gets lower the further you go from TDC. Advanced ignition also raises the combustion temperature, which in turn will make the plug run hotter – and raising the electrode temperature will also lower the voltage requirement.

100

The dimensions of the electrodes also have an effect on the required voltage. Electrodes with square, sharp edges need less voltage and the voltage is also proportional to the diameter of the centre electrode, so thin wire electrodes need a lower voltage. Having less material would mean that the electrode would wear away faster, so special alloys have been developed to counter this, gold-palladium being used in several thin-electrode plugs. NGK introduced a competition version, using a thin-wire electrode but in less exotic materials; it has the voltage advantages, but does wear faster than the gold-palladium type although this is not important in a competition engine and the plug is considerably cheaper.

Increasing the gap between the electrodes raises the voltage requirement, while raising the temperature of the electrodes lowers the voltage requirement.

Finally the electrode temperature will increase with load and speed, and it varies to some extent with the tightening torque of the spark plug (plus the contact made between plug and head, which will be affected by dirt and by the condition of the plug washer). NGK's recommended torque settings for aluminium heads are shown in Table 7.1.

Table 7.1 Spark plug tightening torques

Plug type	thread dia, mm	tightening torque, kg-m (lb ft)
Flat seat,	18	3.5 to 4.0 (25 to 32)
with washer	14	2.5 to 3.0 (18 to 22)
	12	1.5 to 2.0 (10 to 15)
	10	1.0 to 1.2 (7.2 to 8.7)
Conical seat,	18	2.0 to 3.0 (15 to 22)
without washer	14	1.5 to 2.0 (10 to 15)

Types of spark plug

The basic size of spark plugs is determined by the thread diameter and its reach. The common sizes are 10,12, 14 and 18 mm diameter, with reaches of 11.2 mm, 12.7 mm (12.5) and 19 mm (18). The figures in brackets refer to racing plugs; other reaches are available but these are the ones most commonly used on motorcycle engines.

The plugs themselves have various physical characteristics, such as a variety of hexagon sizes, options on the overall length, flat seats or conical seats and the remaining differences are due to the internal construction.

The most important feature is the plug's heat range. This is its ability to conduct heat away from the electrodes; one with a poor heat path will make its electrodes run hotter, in given engine conditions, than one with a good heat path. The heat path is altered by the use of materials and by the length of the central insulator. The many conditions listed above influence the temperature of the plug and the heat range has to match these so that the electrodes stabilize at the optimum temperature.

This is the temperature at which the electrodes are able to clean themselves by burning off deposits from the fuel and oil, and at which the plug's performance in terms of sparking requirement and wear is at its most efficient. At lower temperatures the electrodes will be too cool, deposits will build up, the voltage requirement will increase and eventually the plug will foul. At higher temperatures the electrodes will wear very quickly (ultimately, they'll just melt) and they may get hot enough to cause pre-ignition in the fuel.

There is an interim region where the plug's self-cleaning abilities equal the rate at which deposits form. This depends on the temperature and the air-fuel mixture; a lean mixture will not cause so much fouling and the plug will be able to run at lower temperatures; a rich mixture will need higher temperatures, otherwise it will cause plug fouling. The build-up of deposits raises the voltage requirement until, when the ignition can no longer meet the demand, misfiring occurs.

In the interim region conditions exist which can produce lead fouling (if leaded fuel is used). At ambient temperatures this forms a shiny, yellowish varnish on the centre insulator, which doesn't affect the plug's performance. But in the region of 400°C, this becomes a molten material which is electrically conductive and can cause misfiring.

The heat range of the plug is encoded in its classification number. For example, NGK use numbers, the lower the number the hotter the plug will run in the engine; using a plug with a higher number means that it will run cooler in the same conditions.

In addition to the heat range, there are other design and material differences which make plugs more suitable for certain applications. Some of these are:

Projected nose Similar in construction to the standard plug, but with an extended centre electrode and insulator. This allows the tip of the electrode to respond quickly to engine temperature changes, allowing it to warm up

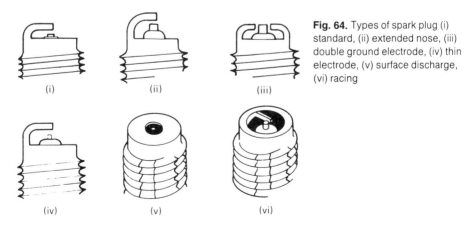

Fig. 64. Types of spark plug (i) standard, (ii) extended nose, (iii) double ground electrode, (iv) thin electrode, (v) surface discharge, (vi) racing

(i)   (ii)   (iii)

(iv)   (v)   (vi)

Fig. 65. Ignition pick-ups on a race-tuned RD350LC

Fig. 66. This RD500 in F1 trim has a racing magneto. Note also the exhaust routing; the fabricated sections of pipe are forced into shape by hydraulic pressure

rapidly when the engine is cold-started, or to maintain its temperature when the engine is idled. This prevents fouling at low speed and cold-start conditions, while being extended into the gas means that the electrodes are cooled when the engine is under load, so the heat range is effectively increased. The plug doesn't have the ability to cope so well with a progressive build-up of deposits, which racing plugs need.

Thin-electrode (see above) Similar in construction to the standard plug, but the use of a thin centre electrode reduces the voltage requirement, improving performance in borderline engine conditions. High wear is countered by special materials, which make the plug expensive.

Dual ground electrode Developed to cope with high temperatures which cause rapid wear; having two gaps virtually doubles its wear resistance.

Surface discharge The ground electrode is a plate which surrounds the centre electrode, leaving an insulator with a very small surface area. Consequently the plug runs very cool and needs to be used with a powerful capacitor discharge ignition which is capable of providing a spark even when the plug has fouled.

Resistor plug The connection through the plug to the centre electrode contains a high resistance, in the order of 5000 ohms, which is meant to suppress radio-frequency noise.

Racing plugs The main difference from the standard plug is that the nose is retracted and the ground electrode is a short, straight attachment to the edge of the threaded section. This is to protect the plug from overheating and mechanical shock. The electrodes may be made of special alloys (nickel or platinum are often used).

Reading spark plugs

From the variety of engine conditions which affect the combustion temperature and the performance of the spark plug, to the differences in the plugs themselves, it is easy to see that 'reading' spark plugs is not going to tell you very much. Not in absolute terms, that is. The colour of the electrodes and the insulator will vary according to the temperature last reached and the deposits will vary with different conditions. So many things can alter those conditions, including the plug itself, that it is impossible to draw any conclusions from one inspection of one plug.

As a comparison, used intelligently, it can have more value. First the plug must have the correct heat range for the conditions; second, the appearance of the plug when the engine is normal must be known; any change in appearance can then be connected with operating changes.

104

A rise in temperature, making the plug lighter in colour, until it is ultimately bleached white, can be caused by mixture changes or by advanced ignition. Where there are conflicting changes – like a rise in temperature caused by exhaust back-pressure, compensated by retarding the ignition or running a richer mixture, the plug readings can be misleading. The plug reading should be taken in conjunction with other symptoms and used to confirm ideas, not to predict them. It sometimes helps to lift the cylinder head and inspect the colour of deposits there and on the piston. A dark colour means cooler running.

Ignition timing

The optimum timing can only be found by experiment, as described above, to use the minimum amount of advance while maintaining torque at each engine speed. As this is critical on most tuned two-strokes, and slight errors will quickly put holes in pistons, it pays to err on the safe (retarded) side and to physically set the ignition position with great accuracy.

The best method is to set the engine position using a dial gauge, use this to mark the flywheel and then confirm the timing when the engine is running using a stroboscope.

1 If the head has a central, vertical plug hole, then a screw-in adaptor can be used to hold the dial gauge. If not, then the dial gauge can only be used in the plug hole in order to locate TDC – all direct measurements will be inaccurate. In this case, remove the head and make up a clamp to hold the gauge securely in the vertical position over the centre of the piston. Keep the barrel bolted to the crankcases (make up a packing piece to go under the head nuts if necessary) as this will (a) prevent the piston accidentally raising the barrel when the engine is turned and (b) it will prevent crankcase pressure from breaking the gasket seal between the cylinder and the case.

2 Locate TDC. This and the timing can be done with a degree disc *if* the disc can be mounted rigidly and concentrically to the crankshaft. To use the disc, make a stop which can be screwed into the plug hole, etc., and positioned so that it stops the piston about half way up the bore. Turn the piston up to it, note the degree reading, turn the engine back the other way until the piston reaches the stop once more and note the degree reading again. BDC and TDC will be half way between the two readings.

3 Set the engine position, to the manufacturer's specified timing or to your own figure. If there is any possibility of backlash in the drive to the breaker/trigger/pulser, then turn the engine too far back and turn it in the normal forward direction up to the timing point.

To use a degree disc, the pointer should be set to zero when the engine is at TDC. The pointer will then indicate degrees before and after TDC; a lot of care has to be taken to align the pointer and to keep it close to the disc so that the reading isn't affected by the parallax displacement if you don't look squarely at the disc.

To use a dial gauge, set the piston at TDC, by rocking the engine back and forth at the point where the needle reverses direction. Turn the sliding scale so that the pointer is at zero, the scale will then read mm before/after TDC. Turn the engine backwards beyond the timing position, then turn it up to the position indicated by the gauge.

4 With the engine in the firing position (for full advance if an advance mechanism is incorporated into the ignition system), mark the flywheel in an appropriate position, and put an equivalent mark on the crankcase casting adjacent to it (the marks may already be there, in which case this procedure simply checks their accuracy). These are to be used with the stroboscope when the engine is running. The ignition trigger has to be set so that it operates at this point:

(a) contact breakers, having been set to the correct gap, should just break electrical contact. This is best checked by connecting an ohm-meter, a battery and lamp, or a battery and buzzer across the points, so that the exact electrical contact can be seen or heard.

(b) with breakerless systems there are some types of trigger which can be seen to function and can be set up accordingly, on others you have to rely on register marks made by the manufacturer, or the measurement of a gap between the moving pieces. This must be set, to be checked by stroboscope timing later.

5 When the engine is built, connect a stroboscope to the plug lead and to an external power source if necessary, run the engine and shine the strobe light on to the flywheel. The mark on the flywheel will indicate where the plug is firing relative to the fixed mark on the crankcase, and this can be checked at all engine speeds. Mark the crankcase level with the strobe-mark, if there is any variation, and re-time the engine to allow for this variation. Repeat the process until the timing is right.

Chapter 8
Carburettors

The theoretical needs of an engine are for an air/fuel mixture of about 14:1 by weight. Most engines give their best power when the proportion of fuel is increased slightly above the chemical ideal, tuned two-strokes go further and require a mixture which is substantially richer. The reason for this is partly that the latent heat of the additional fuel helps to cool the piston and partly because the gas flow follows such violent changes that it is the only way to guarantee enough fuel in the combustion chamber at the appropriate moment.

The theory assumes that the air/fuel mixture is homogenous, that the fuel is broken up into the smallest possible particles which are then distributed evenly throughout the gas. This unfortunately is not the case and it will soon become obvious to anyone who has to deal with two-stroke carburettors that there are three mixture conditions, other than the theoretically ideal: rich, weak and wet.

Irrespective of how much fuel is supplied to the air stream, if it all collects in one liquid blob, it is never going to be possible to ignite it. This is the essence of a 'wet' mixture and it can have the less than endearing habit of making a rich mixture appear weak. It is a condition which is inevitable in an engine like a two-stroke, where violent intake pulses surge through the gas flow and solid interruptions like reed valves and flywheels intercept the high-speed fuel droplets. Much of the fuel must drop out of the airstream, and subsequently evaporate back into it and the machine will only reach a steady state when the evaporation equals the fall-out.

This is beyond the tuner's sight and control. At best, all he can do is measure the flow of fuel into one end of the machine and the production of power out of the other end, – and hope there will be some relationship between the two.

The standard carburation test involves running the engine at constant speed while varying the fuel flow and measuring the load. In the fully rich condition the engine will splutter and four-stroke and make smoke, and as the mixture is leaned off, the load will increase, reach a maximum and then fall off again, with the engine misfiring as the mixture goes fully weak. The result of this test is called a mixture loop, because of the shape that it makes when drawn as a graph of fuel flow versus load. One point of the loop will represent maximum load and the next carb setting along the line in the rich direction is usually the one chosen. This is fairly straightforward, until you find that the jet which gave maximum load at 6,000 rev/min is nothing like the jet required to give maximum power at 9,000 rev/min. The need now is for something to give maximum load at a variety of speeds and this requirement is called a fuel slope – and would be the line drawn by connecting all the optimum points on the mixture loops for each engine speed. To see why one jet cannot cope with all speed conditions and to see what can be done about it involves a little theory about the way in which carburettors work.

To start with, we'll only consider the carburettor on wide open throttle, or WOT. The air being drawn through the carburettor reaches a speed – a fairly high speed – which depends upon the engine speed and its volumetric efficiency. Compared to the air outside the venturi, the high speed air is at a lower pressure, when the kinetic energy goes up, the pressure energy has to fall, as explained in Chapter 4. This pressure difference is used to push fuel through the various jets into the venturi, where the high velocity air smashes it into small droplets and carries it in the general direction of the engine. Now the air flow goes up in – more or less – direct proportion to the engine speed, and so the velocity in the venturi is also proportional to engine speed. But its kinetic energy is proportional to the *square* of its velocity and it is this increase in kinetic energy which causes the *loss* in pressure energy, because the total energy in the gas has to remain constant.

The fuel flow will depend on the pressure drop, and on the size of the jet through which it has to go. If this fuel is adjusted, by changing the jet size,

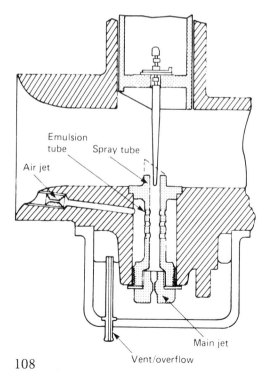

Fig. 67. Essential components of an air-bleeding carburettor

Emulsion tube   Spray tube

Air jet

Main jet

Vent/overflow

Fig. 68. 'Primary choke' type of spray tube used as an alternative to the emulsion tube. The height and shape of the primary choke have a major effect on fuel delivery

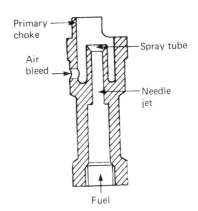

Fig. 69. How the fuel flow at various engine speeds is affected by changes in main jet, air jet and primary choke. Bottom right: a typical tuning sequence. The dotted line (5) represents the fuel delivery which gives optimum carburation through the speed range. Line (1) represents the first attempt at jetting. with a 115 main and 0.8 air jet. This is too rich everywhere. Line (2) is what happens when the main jet is reduced to 110; nearly right at high speed, too rich at low speed. Line (3) is a further main jet reduction to 105; right at low speed, too weak at high speed. Keeping the same 105 main but using a smaller (0.5) air jet, gives line (4) which is close enough to the engine's requirements

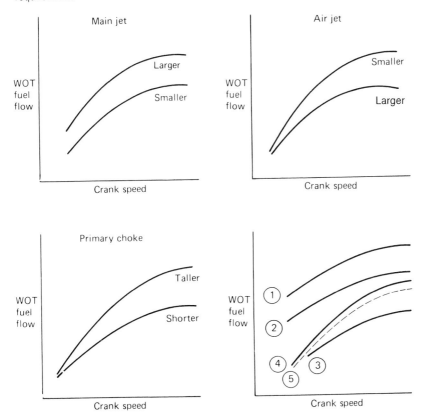

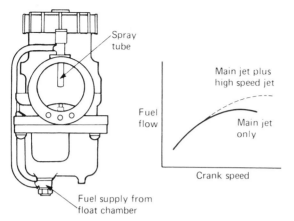

Spray tube

Fuel supply from float chamber

Fuel flow

Main jet plus high speed jet

Main jet only

Crank speed

Fig. 70. Further refinement (i). A high-speed, or power jet, will supply extra fuel in the top half of the rev range

Fig. 71. High speed jet installed on a Mikuni carburettor

until the fuel proportion is correct at one speed, what happens if we then double the engine speed? The air speed will be doubled, but the kinetic energy and the pressure drop will be quadrupled. The fuel flow, assuming the jet can manage it, will be quadrupled as well. Even if other factors, like the engine's volumetric efficiency, friction in the various passages, the discharge coefficient of the jet, etc., modify these calculations, the fuel flow will still rise at a much faster rate than the air flow.

Obviously something has to be done to control the fuel flow over a range of speeds and in slide carburettors this is achieved by bleeding air into the fuel, after it has gone through the jet and before it emerges into the air stream. A small air jet controls the size of this air bleed and like the main air flow, its effectiveness increases with speed and consequently it makes little difference to the fuel flow at low revs, and has its major effect at high revs. There is another advantage in premixing the air and fuel, in that it makes it easier to break up the liquid, improving the atomising process at the venturi.

Fig. 72. Further refinement (ii). Vacuum bleed; there is low pressure in the carb intake at B under conditions of wide throttle opening and high engine speed. Fuel is supplied to the float chamber at A and the only other vent is through B, so chamber pressure is reduced at high engine speed, reducing the fuel flow

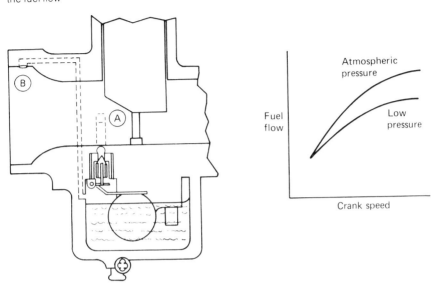

Fig. 73. Further refinement (iii). Second air jet. Sometimes the jetting for wide open throttle is not compatible with that for part-throttle. Here a second air jet is blanked off by the throttle valve until it reaches the wide open position, when a notch or cutaway opens an entry to it. This then reduces the fuel supply at high speed when the throttle is wide open, but doesn't disturb it when the throttle is three-quarters open or less

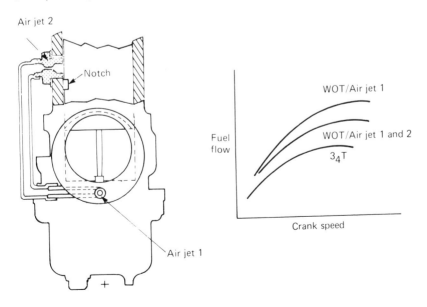

There are two basic methods of introducing the air bleed to the fuel stream: the emulsion tube and the primary choke. In the first, the tube has several drillings made in it and passes through a chamber or well, which is connected to the air jet via a short passage. Air entering the chamber mixes with the fuel and causes a froth of bubbly fluid to emerge at the spray tube. This has the advantage that when the throttle is closed and then opened again, the frothy liquid does not tend to fall back down the tube as quickly as neat liquid might. Also when the engine is idling and little or no air is passing through the air jet, the well fills with fuel, providing an immediate supply of (arbitrarily) rich mixture when the throttle is snapped open. This has a similar effect to an accelerator pump, and tends to prevent flat spots on sudden acceleration.

The primary choke (Fig. 68) consists of a shield around the upstream side of the spray tube and doesn't mix air with the fuel until it has reached the spray tube. It increases the rate of fuel delivery at high engine speeds and is useful on two-strokes with peaky power characteristics, which actually need a large increase of fuel at high revs. The fuel delivery and the degree of atomisation can be varied according to the height and shape of the primary choke. It is also possible to have an emulsion tube and a shield around the spray tube, increasing high speed flow with better atomisation.

There are three other less common devices for tailoring the WOT fuel supplies:

High speed jet   This is a separate fuel jet, which delivers fuel to a gallery in the top of the air venturi, upstream of the throttle valve. A nozzle projects into the air stream from this gallery. Whenever there is enough airflow over the nozzle to cause a pressure drop which will lift fuel from the float chamber, this system (often called a 'power jet') will deliver fuel. Its height and the size of the jet regulate the flow through it, and it will normally come into play at high throttle opening and high engine speed.

High speed air jet   This is a second air bleed whose entrance on the side of the carburettor is blocked by the throttle valve until the carburettor reaches WOT. Then a groove machined in the throttle valve opens the entrance to the air jet. This increases the effect of the main air jet, ie to weaken the mixture at high speed, but it has no effect on the part-throttle settings.

Vacuum bleed   Normally the pressure in the float bowl is atmospheric, that is, constant within daily variations of barometric pressure. However some carburettors do not have the usual external vents to the float bowl but have an internal duct to the top of the venturi, upstream of the throttle valve or into the air box. At low speed and low throttle opening this behaves like a normal vent. At high throttle opening and higher speed, the vent is subject to low pressure because of the gas flow across it. This reduces the pressure inside the float chamber and therefore reduces the pressure across the main jet at high speed. This was used on quite a few of the early 250 and 500 cc Suzuki twins. On machines with air intakes designed to pressurize the air box at high speed similar vents ensure that the float bowl pressure keeps in step with the intake pressure.

Moving on to part-throttle settings, the same criteria apply and the carburation is tuned by the same sort of tests, holding the controls on a fixed quarter-throttle, half-throttle, etc., and changing the relevant jets to get a carburation loop. This is obviously a time-consuming job and it is not made any simpler by the fact that each system overlaps with its neighbours. Usually the tuning sequence is WOT, (main jet, air jet); idle (pilot jet, air/volume flow adjustment, throttle stop); high part-throttle (needle, needle jet); low part-throttle (idle progression, air slide).

Because of the degree of overlap, it will usually be necessary to go over these settings two or three times, neglecting the last one as long as the engine runs tolerably well. The reason for this is that having got all the settings right for constant speed, the carb then has to be tuned for acceleration, to pick up cleanly and respond to the throttle, and to pass without flat spots from one system to the next. The pilot by-passes and the air slide cutaway are largely responsible for the initial pick-up.

Main jet

There are five or six different types of main jet available, some with different thread sizes, some with hex heads and some with round heads. Even so it's still possible to be confused between two or three types but, as long as you know what the differences are, it's easy to stick with the same type.

The first difference is in the construction of the jet – the orifice – which will have a streamlined entry. On some types, called reverse flow, the orifice is set further into the body of the jet and the larger sizes of this kind flow a lot more fuel than the same size of the other types.

Sizing brings us on to the other problem. Some are stamped with the jet diameter (in mm), others with a flow rate in cc/min. As the two ranges of numbers overlap, it's easy to confuse them.

Finally there are plastic jets as well as the traditional brass ones, and although the nominal sizes may be the same, the flow rates may be slightly different. It doesn't matter if you stick with the same kind, then a 105 will always flow more than a 102.5, but if you mix the two kinds, this may no longer apply.

While the main jet is selected for power in the three-quarter to WOT position, it will have a diminishing effect right down to quarter throttle, or maybe further. This is why the main jet has to be selected first.

The usual procedure is to aim deliberately rich and work down until the carburation is close, go over the other systems and then have another go at the main jet. Depending on the facilities available, the best main can be chosen for power at the bottom of the speed range, and the air jet used to tailor fuel flow at the top end, or you can start at high speed and trim the flow at low speed – it doesn't matter, as long as you can recognise signs of weak running before it gets dangerously weak. The warning signs are: a rise in temperature, knocking, power fade, you may hear the edge go out of the exhaust note, there may even be a slight misfire, overheated spark plugs. If you're riding the bike you'll probably only notice the power fade – you'll feel the bike slow fractionally, the rev counter may drop a fraction, you may be able to detect the change in exhaust note. On a low powered bike you may also hear any knock. It is unlikely that the coolant temperature gauge would pick up the increase quickly enough; a thermo-couple connected to one of the spark plugs is more likely to give a more rapid warning. A full race motor won't stand more than a few seconds of really weak running, and some machines are so sensitive to carburation that they will go through a full loop in only three or four main jet sizes

This sort of sensitivity is usually a sign that something else is wrong – a mismatch between intake and exhaust, too high combustion temperature etc. but this doesn't mean that it will be easy to find or to cure.

Because of the critical nature of the carburation, it is essential that everything else, particularly ignition timing, is correct before attempting to get the mixture right.

Air jet

As explained above, this is used to modify the main jet's fuel slope to get the best fuelling over a wide speed range. On most production machine carburettors, it is not a replaceable jet but is usually a simple drilling or a brass jet pressed into a drilling. Consequently, if it is necessary to make it adjustable the passageway must be tapped with a thread to take new jets. Air jet kits are available for most carburettors.

If there isn't enough scope with the air jet, then the choice is to get the best fuelling at peak revs using larger fuel jets/smaller air jets and let it run rich at lower speeds, or to fit a high speed jet. In this case, the fuelling can be tuned for lower speeds and what would be a weak mixture at high speed is corrected by the high speed jet. Some GP bikes have two high speed jets, arranged to come in progressively. Some of the larger Mikuni carburettors have high speed jets fitted or have provision for them and there are aftermarket kits to fit them to other carbs.

One other possibility is to use a different spray tube, with a shield like the primary choke, in order to gain some high speed richness, plus better atomisation. This often means that the main jet can be reduced by several sizes and makes it easier to get the right fuel slope.

Pilot jet

This and the cold start jet have to provide for cold starting, idling and the progression off idle. Usually the stock jet will not need to be changed, but the mixture control setting will and the accuracy here depends upon the sensitivity of the operator in getting the highest engine speed on the lowest throttle stop setting. It may need to be richened beyond the optimum in order to avoid a flat spot as the throttle is opened.

The pilot outlet will be the one furthest downstream from the throttle plate; there will also be a by-pass under the back edge of the throttle, to bring in more fuel as the valve is lifted. It may be necessary to enlarge this slightly, if there is access to it, by using a hand-held drill bit and carefully opening the drilling to the next size up. The alternative may be to have the idle abnormally rich.

Intake cross-connections, or chambers connected into the intake can make large differences to the idle requirements and if the idle is particularly rough it would be worth experimenting with something along these lines.

Air slide

Throttle valves with different cutaways are available, or the existing cutaway can be enlarged. It has most effect in the ⅛ to ¼ throttle region, particularly in providing a smooth transition on to the needle system. This is one area which is best tested on the track, as it is pick-up and throttle response from a variety of speeds which is the important thing.

Needle and needle jet

This controls the part-throttle fuelling from quarter throttle to three-quarter throttle, plus some overlap at each end. It is affected by the main jet, air jet and the pilot settings, but in this wide region the needle is the dominant factor.

Mikuni code their needle jets, using a letter and a number. The letter gives the basic bore size in mm, in increments of 0.05 mm; so N is 2.55, O is 2.60, P is 2.65 and so on. The numbers represent an increment of 0.005 mm. So N–0 is 2.550; then N–1 is 2.555; and N–2 is 2.560 etc.

Fig. 74. Major dimensions of Mikuni needles are coded in the part number: a – overall length; b – length of parallel section; c – length to second taper; A – angle of first taper; B – angle of second taper. The fitting grooves are numbered 1 to 5, starting from the top

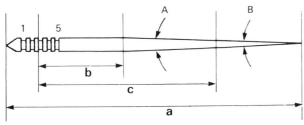

The needles are also coded but this is a little more complicated because they usually have two tapers. The full code can be illustrated by an example: 5GN36-3 as used on the RD125LC Yamaha:

(a) the first number (5) indicates the overall length, between 50 and 60 mm, in this case. A 4 would indicate 40 to 50 mm.
(b) the first letter (G) indicates the angle of the upper taper.
(c) the second letter indicates the angle of the lower taper (N).

The letter code uses every letter in the alphabet, starting with A which represents 0 degree 15 minutes, and going up in 15 minute (ie quarter-degree) intervals. From this, the RD's needle has a top taper of 1° 45' and a bottom taper of 3° 30'.
(d) the next number (36) is the manufacturer's lot number.
(e) the last number is the groove in which the needle is located, in this case the third, counting from the top.

Needle adjustment is made basically by moving the needle clip to a higher or lower groove. A lower groove raises the needle in its fitted position and therefore richens the mixture.

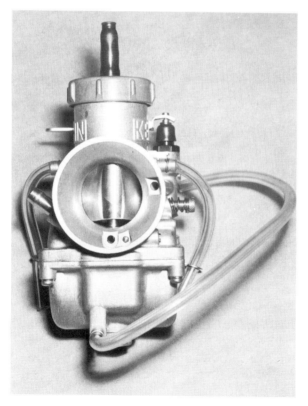

Fig. 75. Keihin carburettor. Note polished air slide. The air bleed just above the 3 o'clock position supplies the cold-start jet

If the needle cannot be adjusted to give the right mixture then a new needle may be necessary. To change the mixture in the lower half of the throttle opening, change the top taper. To change the mixture from half throttle upwards, change the lower taper. A large taper, ie a letter further down the alphabet, gives a progressively richer mixture as the throttle is opened further.

To alter the mixture strength all the way through, a new needle jet may be called for.

One problem can arise here. The main jet must be smaller than the needle jet plus needle on full lift. The use of a large needle and a large main jet may prevent this; the needle jet will then control WOT fuelling and there will be no response to main jet changes. A simple test for this condition is to run the engine without the main jet fitted.

Float chamber

While the float height is not usually critical, it has to be set somewhere, so it may as well be at the height that the manufacturer recommends. Most carburettors intended for motorcycles seem very sensitive to flooding problems, either through dirt in the fuel which makes the float valve stick, or because the needle valve cannot seal against a head of more than a few inches.

Consequently it is a good plan to make sure that the needle valve and its seat are in perfect condition and that there is a filter in the fuel line. On some carburettors, e.g. Mikuni, it is possible to use a smaller needle valve, which has a higher seat pressure. This may be necessary if a fuel pump is used, or if the fuel is to be supplied at a greater head of pressure.

Types of carburettor

Because of installation problems, the choice on most two-strokes is limited. Mikuni is the most popular make, followed by the larger Amal. There is a flat-slide Mikuni which is more compact than the round-slide version and is supposed to give better pick-up from the different slide shape. Tuning is essentially the same.

As the length of the intake tract is important, it may be necessary to add an intake trumpet to the carburettor, as the easiest way of altering the overall length. It may also be necessary to change the length or to alter the position of the carburettor to avoid unwanted pulse effects which would otherwise upset the carburation.

The size of the carb's bore is an important factor in its effect on both the power and the carburation. The carburettor must be able to flow enough air without restricting the engine, yet if it is too big it will become difficult to tune and it will shorten the usable rev range, because it will not work well at lower speeds.

A bore size of 22 mm will give 18 bhp without difficulty, while 30 mm carburettors can handle 30 horsepower and these proportions are close to the limit for flexibility where road bikes are concerned.

Associated problems

While the air/fuel flow controls the engine's power, the fuel is also used to cool the engine, which is why unusually rich mixtures are commonplace in competition two-strokes. When the best power mixture is used, the initial power delivery is high, but it fades after a few minutes, eventually losing 10 to 20 per cent of the original output for an air-cooled engine, and about half this amount for a liquid-cooled machine. Increasing the richness of the mixture will knock a couple of horsepower off the first reading, but it will reduce the power fade to something like 5 to 10 per cent after several minutes' running, and less than this for a liquid-cooled motor.

When setting up carburettors on a dynamometer using fuel flow meters (or even exhaust gas analysis) the usual procedure is to emulate the SFC of the original engine (see Chapter 10) and to make the mixture a little richer in order to minimize the risk of detonation. Quite often, despite an increase in SFC or the presence of unburnt fuel in the exhaust which normally indicate a rich mixture, the engine speed or load goes up. It is hard to tell what is happening to the extra fuel. It could simply be cooling the piston. Alternatively, the process of drawing the fuel/air stream through the crankcase and scavenge ports means that the heavier particles of fuel are likely to lag behind the main flow; the mixture strength of the gas entering the cylinder is likely to vary. It may be that the weaker part of the mixture is trapped in the cylinder, while the richer portion ends up in the stream which 'short circuits' across to the exhaust port. Extra fuel would, in this case, ensure that the trapped gas had the right fuel/air mixture. Either way, unburnt fuel will appear in the exhaust system. This process sometimes continues, with larger main jets simply giving more power, not extra richness. It may finally stabilize with an SFC as high as 0.9pt/hp-h. A problem which can then occur when the main jet is made substantially bigger is that it becomes bigger than the effective area of the needle jet and needle, with the result that full throttle fuel flow is controlled by the needle jet and not the main jet.

Further increases in main jet size make no difference. This condition can be tested by running the engine with the main jet(s) removed. One solution is to fit a larger needle jet although this then requires the complication of finding a suitable needle, with all the extra part-load tests that this involves. Another solution is to make modifications which will cause the same-size main jet to flow more fuel – usually by reducing the air jet size or by using a needle jet with a screen in front of the nozzle.

There are other problems which are not of the carburettor's making, but which either show up in the carburation or can be cured by carburettor changes.

Detonation can be held at bay by making the mixture richer.

Reed valves which do not seat properly, or which flutter, will cause starting and idling problems because the crankcase gases can blow back at low engine speeds.

On very powerful engines, the original fuel tap and fuel lines may not be adequate and may make the machine run weak at peak power, simply through fuel starvation. A similar problem may be caused by inadequate air vents in the fuel tank.

On machines which are changed to run on a fuel/oil mix, the carburation will have to be re-jetted to accommodate the extra weight of the oil carried by the fuel (see lubrication, Chapter 9).

Chapter 9
Lubrication

The extremes of the conditions inside a two-stroke engine put severe demands on its lubrication system. After putting in enough oil to prevent seizure under full load, this oil then has to be removed, either by burning or by loss through the exhaust. If it is not removed it will form deposits and varnishes which, in turn, lead to combustion problems (plug fouling, pre-ignition), ring-sticking (which causes blow-by and yet more piston burning/varnishing), and deposits on working surfaces which will increase friction and reduce heat transfer. The presence of too much oil can evidently be as hazardous as the lack of oil.

In a roadster, where the long-term build-up of deposits is more important and where a degree of corrosion protection is also necessary, there are different oil delivery requirements from a competition machine. For roadsters tuned to less than 200 bhp/litre, the throttle-regulated pump is probably the best system, as it provides better control of lubrication in the non-extreme conditions and is generally a lot more convenient to use. At higher levels of performance, or in competition engines, where long-term factors can be ignored, there is more to be said for pre-mixed fuel and oil. In order to cope with the worst conditions, pre-mix tends to deliver too much oil at other times and the deposits which this causes need cleaning quite frequently.

If there is a problem with deposits building up, then the ratio of oil to fuel could be reduced – it really is a matter of experiment to find the best level, consistent with maximum power production and sufficient engine protection. As too much oil will also cause problems, it can be important to find this optimum level. The manufacturers of competition engines usually recommend a fuel to oil ratio of 20 or 24:1, some going up to 30 or 40:1 or even higher. The oil companies will not usually make general recommendations, saying things like 'up to 40:1' – but are usually quite helpful with recommendations for specific applications.

Ultimately the choice lies with the operator and the only precise method is a series of dyno tests to check the effect on power, plus endurance tests or track experience to determine the level of engine protection. An alternative to changing the fuel/oil ratio is to use a different type of oil, which may be the answer to certain problems where there is not enough difference between too little oil causing seizure and too much oil causing fouling.

Two-stroke oil intended for roadsters usually has an SAE30 viscosity base, and can be mineral or synthetic (or a combination of both) but its development and additives will be aimed towards low ash content, low smoke emissions, anti-oxydants and so on. It may not have the anti-seize properties which a tuned engine will need. Heavier oils, which have better load-bearing properties, will tend to be SAE40 and above which may be too thick for injector pumps to manage.

There is a conflicting demand here. The oil is served up to the engine with no pre-heating, in a random mist feed. It needs to have lubricating properties – in which the load-bearing capacity is important, and this tends to be better with high-viscosity oils. To protect the engine in extreme conditions, anti-seize properties are also essential and these depend on the lubricant's oiliness plus EP additives and the like. The bad news is that high viscosity oils and this type of additive usually cause worse deposits and varnishing.

The oil film, particularly between the piston and cylinder, also has to transfer heat. In this role, low viscosity oils are usually the best.

Oils intended for competition use fall into two broad categories. There is castor oil and the mineral/synthetic types. Castor oil still has the best anti-seize properties but it has many disadvantages. It cannot be mixed with mineral oils, it causes heavy varnish deposits requiring frequent engine overhauls, its miscibility with petrol is not the best so the fuel and oil should only mixed immediately before use.

Specialised synthetic and part-synthetic oils for competition two-strokes have been formulated, mainly for use in pre-mix systems, but some can be used with a separate pump. These would appear to be the best option for most high-output engines, although they still require a lot of experimental work to determine the best type for a particular application, and the best mixture ratio.

Lubrication problems can also be caused by engine clearances which do not match the needs of the oil. The piston skirt clearance is a critical factor, as too great a clearance will cause varnishing which will lead to overheating, increased frictional losses and eventually to seizure. A clearance which is too small will not allow enough lubrication, again with the risk of rapid seizure.

Deposits in the ring grooves, causing sticking and blow-by, will be aggravated if the ring/groove clearance is too great.

Carrying oil in the fuel will also affect the carburation, as it will increase the density of the fuel and reduce its flow through a given jet size. If there is a volume change when the oil is added to the fuel, indicating that it is not completely soluble, then the jets will have to be increased even further. An increase in jet capacity of something up to 10 or 15 per cent will probably be called for, but it is likely to need a complete re-tune with the added complication that a rich mixture will also produce over-oiling and perhaps oil fouling. The presence of oil in the fuel will also affect its octane rating and may have some effect on combustion as well. Some oils contain additives which either stabilize or increase the octane rating of the fuel. They also make the engine noticeably smoother.

The theory is that when the fuel/oil mixture gets into the crankcase, the lighter fuel fractions evaporate, leaving oil mist in the form of droplets which settle on the crankcase surfaces. Other smaller droplets obviously get carried into the cylinder with the scavenge gas containing the heavier fuel fractions, which still contain dissolved oil. Consequently a fair proportion of the oil will either be deposited in the upper cylinder or will be burnt with the fuel.

So far (1993) there is no API classification for high performance two-stroke oil, so there are no standards other than reputation and price by which to judge the available oils. Unfortunately some oils are packaged and priced to resemble high quality, competition lubricants when their performance is no better than any cheap oil available on a garage forecourt.

Where tuned engines are concerned, the main properties needed in an oil are load-bearing capacity and the ability to prevent scuffing or seizure; the oil also needs to be miscible with the fuel or to be thin enough to be used with an injector pump (ideally SAE 20). The oil should also have additives to prevent carbon deposits, varnishing, plug fouling, to resist oxidation in the oil, to prevent corrosion in the engine, and so on. It also helps if the presence of the oil does not spoil combustion or reduce the fuel's octane rating. Some of the long-term properties – like the build-up of varnishes or carbon deposits – are less important in competition engines which are overhauled frequently.

The oil's load-bearing ability can be measured by wear tests on ZN or Falex test rigs. In these tests a static and a spinning component are pressed

Table 9.1 Applications for various types of two-stroke oil, based on ZN and Falex wear tests

Engine performance	Oil	Type
Over 240 bhp/litre	Shellsport R	Castor based
	Putoline Racing	Castor based
	Castrol A747	Synthetic
	Shellsport S	Synthetic
	Motul 400 (later 800)	Synthetic
	Silkolene Pro 2	Synthetic
200 to 240 bhp/litre	Putoline MX5	Premix only, synthetic
	Motul 300 (later 600)	Synthetic
	Castrol A545	Synthetic
	Amsoil 100:1	Premix only, synthetic
	Bel Ray MC1+	Synthetic
150 to 200 bhp/litre	Putoline Super TT	Synthetic
	Silkolene Comp 2	Synthetic
	Rock Oil K2	Synthetic, premix only
	Bel Ray Si7	Synthetic

together with a known force and the amount of wear is measured after a given time. In the ZN test, the force divided by the area of the wear scar shows the pressure which the film of oil had to support; the appearance of the wear scar (varying from smooth to being badly scored and picking up) also shows how much scuff resistance the oil has.

This test shows the properties of vegetable-based oils very clearly; their film strength is high but even when the oil is totally overloaded, it resists seizure and scoring of the metal surface. As the major requirements of a two-stroke oil are its load-bearing and anti-seize properties, these tests are quite relevant and put the oils into groups which can be matched to various levels of engine development. Of course, there are still other considerations, such as the oil's effects on knock and combustion, build-up of varnish, protection against corrosion, deterioration of the oil itself, etc., which must be added to the oil's load-carrying performance. But, until there is a standard set for two-stroke oils, wear tests at least establish an order of lubricants which are known to be reliable. Oils which have performed well in Falex and ZN wear tests are listed in Table 9.1.

Chapter 10
Development, testing and preparation

An engine may be thought of as a series of processes, each one dependent on the previous one in terms of flow, efficiency, etc. The final output will be governed by one of these processes, if it imposes a greater restriction than the others. An obvious example is the throttle; when it is closed it is the major restricting factor and prevents all subsequent processes from working to the full.

Development then becomes a series of steps to locate the current restriction and reduce it until the engine's output becomes dependent upon some other part. This will continue until mechanical reliability becomes the limiting factor, or until all components are exerting an equal influence on the power production.

Knowing the physical limitations of the engine it may be possible to set a realistic target, based on a safe maximum speed and a load which can also be predicted. Port timing to match the maximum speed can be estimated (see Chapter 4) and maximum port widths are dictated largely by the bore size. Hence a maximum time-area can be found (see Appendix). This can be related to the time-area of the engine in standard form and, as long as components such as the carburettor, exhaust system and ignition can all be made to match the new state of tune, then a reasonable forecast can be made for the power output. If the new time-area at peak speed coincides with the original time-area at, say 9,000 rev/min, then it is assumed that the air flow and torque reached at 9,000 can be maintained at the new speed.

Comparison of the original power characteristics with the time-area and the carburettor size will also show where the engine runs inefficiently and these characteristics may be used to predict the lower limit of the useful power band.

This gives a realistic target, and the development which follows would then take the steps outlined at the end of Chapter 4. At each step it is obviously necessary to determine what is restricting the engine (unless it has immediately given the target output . . .) and concentrate further efforts there.

Symptoms given by the engine often indicate the type of restriction, which can then be confirmed by simple experiments. If the airflow is

Table 10.1 Yamaha 250 cc twin, development history

Model	TD1A/B	TD1C	TD2	TD2B	TD3	TZ250A	C	D/E	F	G	H/J	K/L
Stroke, mm	50	50	50	50	54	54	54	54	54	54	50	50
Bore, mm	56	56	56	56	54	54	54	54	54	54	56	56
Displacement, cc × 2	123.2	123.2	123.2	123.2	123.7	123.7	123.7	123.7	123.7	123.7	123.2	123.2
piston area mm² × 2	2463	2463	2463	2463	2290	2290	2290	2290	2290	2290	2463	2463
Output, bhp	32–35	40	44	47	49	51	52	53	53	55	57	59
Crank speed, r/min	9500	10 000	10 500	11 000	10 500	10 500	10 500	10 500	10 500	11 500	11 000	11 000
Exhaust opens °ATDC	82	82	80	80	79	79	83	79	79	81	79	79
Exh width, % of bore	60.7	60.7	62.5	66.1	68.5	68.5	68.5	68.5	73.2	72.2	73.2	73.2
Exh sp. time-area s-sq mm/cc × 10⁻³ at:												
9000	16.42	16.42	–	–	–	–	–	–	–	–	–	–
9500	**15.56**	**15.56**	16.99	17.96	20.06	20.06	17.99	–	21.42	–	–	–
10000	14.78	**14.78**	16.14	17.06	19.06	19.06	17.09	19.06	20.35	19.04	19.24	19.24
10500	–	–	**15.37**	16.25	**18.15**	**18.15**	**16.28**	**18.15**	**19.38**	18.13	18.32	18.32
11000	–	–	–	**15.51**	17.13	–	15.54	–	–	17.31	**17.49**	**17.49**
11500	–	–	–	–	–	–	–	–	–	**16.55**	16.73	16.73
Scavenge opens °ATDC	114 tc 120	114 to 120	114 to 120	111 to 117	109	109	109	109	109	114	116	115
Scavenge: sp. time-area at peak speed s-sq mm/cc × 10⁻³	5.9	7.13	9.14	10.04	13.68	14.4	15.48	15.48	15.48	11.08	10.82	12.42

TD1A: Anodized bores. *TD1B:* porous chrome plated bores; new head; piston skirt cut away to increase intake duration from 140° to 180°; exhaust duration increased by 2 mm cutaway in piston crown. New exhaust system.

TD1C: 27 mm carb; intake port lowered by 4 mm; thinner piston ring (1.2 mm); boost scavenge port; multi-taper exhaust system. *TD2:* five-port engine (main scavenge angled up at 15° to horizontal, secondary port angled horizontally); 30 mm carb; 1 mm piston ring; new head. *TD3:* bigger intake; 34 mm carb; new scavenge port shapes, bigger secondary ports; squish clearance reduced from 1.2 mm to zero; horizontally split crankcases, 6-speed gearbox, Femsa ignition. *TZ250A/B:* liquid cooled; 54 mm stroke retained to share components with TZ350 (64 × 54 mm); bigger main scavenge ports. *TZ250C:* wider intake port; wider scavenge ports. *TZ250D/E:* higher compression. *TZ250F:* wider intake port. *TZ250G:* bigger intake; main scavenge port narrower, secondary port wider. *TZ250H/J:* intake wider, secondary scavenge port wider; power valve in exhaust; transmission jackshaft to allow engine to run backwards, taking thrust off the rear cylinder wall and allowing still larger intake port. *TZ250K/L:* wider intake; larger main scavenge port. The effect of the power valve can be seen clearly from the specific time-area figures: with the valve up, the time-area at 8500 would be 0.023-sq mm/cc – ample time for the loss of fresh charge into the exhaust. The power valve reduces this to just under 0.014-sq mm/cc – more or less the same as a tuned roadster might have.

Base data taken from *Yamaha two-stroke twins* by Colin MacKellar (Osprey), which gives a detailed history of Yamaha's twin cylinder machines.

obstructed upstream of the carburettor then attempts to increase it will simply result in an increase in the mixture richness, because there will be a greater pressure drop across the carburettor. A similar restriction downstream of the carburettor will defeat attempts to raise the air flow, but the carburation will not change although the mixture may become 'wet'. Experiments with the intake time-area or the reed valve should then make significant differences.

The timing of the ports is related mainly to engine speed, while, to provide the necessary air-flow at that speed, the time-area has to be sufficient. Deficiencies here, or a narrow power band, can be attributed to lack of time-area or a mismatch between the timing of different ports.

A ragged power peak – or perhaps even several minor peaks, suggests that either the carburation or the ignition is unstable, either because there is a fault or because it simply cannot cope with the demand.

A physical restriction in the exhaust will obstruct gas flow and will also cause a sharp rise in combustion temperature. A dimensional mismatch is less easy to define. The exhaust has a major effect on the power characteristics, but as the power level is increased, the engine will become less tractable and it is more difficult to tell whether the width and shape of the power band is just the natural consequence of that state of tune or whether the exhaust is making a mismatch somewhere.

Measuring performance is the hard part, unless there is access to a dynamometer and fuel flow meters, first because it is hard to distinguish small changes while riding and second because the weather conditions can easily change more than the bike. There are also conditions which can be misleading, for example the two-stroke's habit of using petrol to cool the engine. The mixture strength can often appear to be right, the combustion good, the power stable and the specific fuel consumption in the safe region, around 0.6 pt/hp-h or a shade higher. But if the main jet is then increased – perhaps to make sure that the mixture strength is safe – the motor doesn't run rich at all. On the dyno you would see an increase in speed because the extra fuel will have cooled the piston and reduced friction by a small amount. If the higher speed suits the porting, exhaust, etc., just by chance, the engine will flow more air and make more power. The mixture and the SFC will stay the same! Or if you're unlucky, it may even run weak . . .

The only answer is not to take anything for granted, check the results and, preferably go too far – in the above example, keep increasing the main jets until it makes symptoms of running rich, and then go back to the optimum setting.

Dynamometer testing is by far the easiest way to set up engines and to optimize the fuelling, ignition timing etc., but it is important to realise that the dyno is only a tool and is therefore only as good as its operator, and that dynos themselves differ in design, giving certain advantages and disadvantages for different applications. Good operators can work to internationally recognized standards and can obtain results which can be compared between one dyno and another – but this shouldn't be taken for granted. If tests have to be run at different locations or in very different ambient conditions then a calibration run should be made, with the engine in the same condition as it was for the previous test, rather then rely on air density correction factors.

There are four main types of dyno.

1. *Water brake* (e.g. Heenan & Froude DPX, Schenck U1-16H, Superflow 800 and 901). This type has a rotor, driven by the engine, which turns inside a housing to which water is delivered under pressure. The rotor has cells angled against the direction of rotation and the housing has similar, opposing cells so that the motion of the rotor drives the water into the housing and tries to turn it. The whole thing is pivoted on a sturdy frame (the 'bed', hence test-bed) so that even friction in the glands and journals supporting the rotor also tries to turn the housing – and therefore gets measured: there are no losses. The whole thing is restrained by a torque arm which carries a weight and a spring balance or an electrical load cell. Once the engine is in a steady state and the dyno is balanced, the engine torque will lift the weight and give a reading on the spring scale – the torque at the dyno is the sum of these two weights multiplied by the length of the torque arm horizontally from the centre of the rotor. The horsepower developed is $2\pi NT/33000$ where N is the dyno shaft speed in rpm and T is the measured torque in lb-ft. (Note the crankshaft speed will be NG where G is the reduction ratio between crankshaft and dyno shaft; the crankshaft torque will be T/G.)

The dyno is controlled by sluice valves which slide in between the rotor and the housing cells, blanking off a portion of the cells and reducing the efficiency of power transmission. Depending on the size of the rotor and its geometry, the dyno has clearly defined maximum and minimum speeds and loads – each type comes with a chart showing these parameters and obviously the engine's load and speed range must lie within what the dyno is capable of measuring. This can be altered to some degree by changing the gear ratio between the engine and the dyno, but any increase in speed is accompanied by a proportionate decrease in torque and vice versa, so this kind of adjustment is limited in its effect.

The drive can be taken by chain from the gearbox sprocket or a spline adaptor which slides over the gearbox output shaft and is coupled to the dyno via a shaft and universal joint. It could also be taken via a roller from the rear tyre.

With the engine running on wide open throttle, the valves are opened or closed to bring the engine to the required speed and once it has stabilized the speed and load are recorded. The Superflow types have computer control in which the minimum and maximum speeds are programmed, along with the speed interval and the time for which the engine is to be held at each speed. The full run is then made automatically, the data being recorded by the computer. This has the advantage that it can be run quickly, simulating transient conditions as the machine accelerates, or it can be held at each speed station for as long as the operator requires (e.g. long enough to let fuel flow meters stabilize).

There can be differences in power readings caused by different techniques. Taking a power reading very rapidly (the so-called 'flash' reading) usually gives a figure significantly higher than the stable reading. Standards for engine testing usually specify the conditions, for example ISO4106 requires that the engine be run at steady load, speed and temperature levels for 30 seconds before the reading is valid. In practice it doesn't make a lot of difference as long as the operator is aware of the potential differences and as long as the results are not compared to those from different test methods.

If the fuel flow is measured (in pt/h or lb/h) then the specific fuel consumption (SFC) is fuel flow/bhp, in units of pt/hp-h or lb/hp-h.

2. *Eddy current dyno* (e.g. Froude G4, Bosch FLA203 – see below). Here the dyno drives an electrical generator whose output is fed back into the field coil to oppose the engine's motion. It can be mounted in the swinging frame bed, like the water brake, or be part of an inertia dyno (see below). The control is by altering the field coil current to change the load and so bring the engine to the required speed; it can be manual, semi-automatic in which the machine will hold either the load or the speed constant at a pre-set value while the other is monitored, or fully automatic in which a programmed load/speed cycle is fed in, usually to simulate road loads for testing engines in transient conditions. The test methods are the same as those for the water brake.

3. *DC machine* (e.g. Laurence-Scott). A similar arrangement to the eddy current dyno, except that a dc machine is attached to the engine and can be used as a motor to drive the engine or as a generator to be driven by the engine. Fitted in a swinging frame bed, the drive etc., is the same as for the water brake.

4. *Inertia dyno* (e.g. Dynojet 100, Bosch LPS002 and FLA203). In this type the drive is taken from a large diameter roller which is driven by the back wheel of the motorcycle. It is connected to a very heavy flywheel (around one ton in the case of the Bosch types, half a ton on the Dynojet) and as the inertia of the flywheel is known, the torque required to accelerate it at a given rate can be calculated. A speed sensor on the flywheel provides data which is sampled by a computer and stored against values from its real-time clock. Software then calculates the acceleration, torque and horsepower, which are stored against roller speed or engine speed (from an ignition sensor) in a computer file.

The Bosch types differ from the Dynojet by having twin rollers. The loaded one is the front one, while the rear is relatively light and turns freely. Speed sensors are fitted to both and if the machine senses too much slip at the loaded roller (because the two speeds will differ) then it warns the operator. The LPS type is a pure inertia dyno, while the FLA also incorporates an eddy current dyno built in to the flywheel. This can be used as an inertia type or power can be fed into the generator to build in an extra load. This can be related to speed, in order to simulate road load conditions.

As the flywheel continues to drive the wheel after the engine has reached peak speed and is shut down, the speed sensor can log its deceleration and the overrun torque can be calculated. If the gearbox can be put into neutral, this will measure the transmission losses; lifting the clutch will give a similar figure, but will obviously include a certain amount of clutch drag and some clutch release bearings do not take kindly to this kind of action. However, where results are quoted as 'corrected to output at the crankshaft' then these 'losses' have been added to the measured figures.

The advantage of this type of dyno is the speed and ease with which tests can be made, as a complete machine can be fitted to the dyno in seconds with no mechanical work or modifications. Given a knowledgeable and careful operator the results can compare very closely with figures taken from other types of dyno. Pure inertia types cannot hold the engine at a steady speed, which is a disadvantage if it is necessary to let certain things stabilize (fuel flow meters can take several seconds). The other disadvantage is that rollers can be hard on tyres because they deform them more than flat ground and, if the machine is strapped down too tightly, it is possible to exceed the tyre's load/speed index which may damage the carcass internally. Twin roller dynos are more severe than single roller types because the tyre is deformed twice in quick succession and extensive testing may produce fatigue failure. If a long series of tests is planned it is as well to use a worn tyre which will not be used again on the road/track.

Inertia dynos, and eddy current types which have a heavy rotor, will continue to drive the engine if it fails mechanically – possibly increasing any damage – until the operator disengages the clutch. Water brakes simply stop when engine power is removed.

All tests are comparative, and as long as you can duplicate the conditions and change just one thing at a time you should be able to make sense of the results. You need a baseline – if it is power then you can use the power curves quoted by the manufacturer or by magazines. The manufacturer's power figures will probably be gently inflated, but the characteristics will be accurate – if they say it peaks at 9,000 rpm, then it will. The other thing to consider here is that the rev counter fitted to the bike will probably have some error; they rarely read low but can be up to 10 per cent high at peak revs. Again, it doesn't matter – you're not interested in absolutes, only in changes – as long as you don't try to compare your tacho readings with someone else's.

It is possible to measure small changes at the track, or even on a suitable piece of road. The easiest way is a top-gear roll-on; opening the engine up to wide-open throttle from a particular engine speed and timing it between two points, or measuring the time interval for it to increase by 500 or 1,000 rpm. As the power output and the speed goes up this gets progressively more tricky. It may be necessary to repeat the runs several times to get consistent results.

There is another limitation, caused by letting the motor run in transient conditions and not holding it in one place long enough to reach a steady state. An example will show how the tests can be arranged – and their limitations.

I once set up a friend's production racer using a stopwatch and a track – just to see if it could be done. The engine was put very carefully to stock settings, ignition timing, plugs, carburation – everything was prepared in as-new condition. To save time in setting the ignition with a dial gauge at the track, we made several settings in the workshop and marked the backplate in three or four divisions of advance and retard. The throttle was also marked so that the carburation could be checked in several part-throttle conditions, as well as wide open.

At the track we tried various markers and found that something about 150 yards apart was the optimum for one person to time using a stopwatch – as long as the rider followed exactly the same approach line each time. The bike was timed over four or five runs – more if we got any spurious results – at each of the throttle settings. We then proceeded to alter things, one at a time, and repeated the test runs. Thus, using the ignition timing which gave the quickest time, we went on to change the main jets, and then needle settings, finally going back to recheck the ignition and last of all the main jets one final time. It took all day and we had to make enough changes for the performance to deteriorate, so that we could select a 'best' setting and return to that. Fortunately there was a loop in the track, so the rider didn't have to do a full lap. It took an effort on the rider's part, to approach the marker at exactly the same speed, on the same line, to open the throttle at precisely the right point and to hold the same posture (changing any one of these would change the time as much, or more, than the changes we were making to the bike.)

The surprise was that we could get consistent results and that we could measure changes. During the course of the day we saw quite significant changes in times – in the order of 15 per cent – and at the end, when we put the bike to its 'best' settings, we reproduced the same times. The stopwatch said it was quicker, the bike's tacho agreed, the rider said it felt noticeably crisper and it even sounded better. It was a long, tedious job but it had worked.

That weekend the bike was entered in a club race and, during practice, it had easily outperformed a couple of bikes which it had previously been equal to. In the race it pulled clear of its usual opposition, and halfway through it melted the tops of both pistons. It says something for our painstaking care that both pistons were equally damaged – usually one goes a long time before the other.

There were two reasons for the failure. First we'd set the bike in transient operating conditions – it was always accelerating from one speed to a slightly higher one. Second, it was only held wide open for a few seconds. By not running at a constant speed and by not holding the power on for a reasonable length of time, we had produced slightly lean mixtures, advanced ignition timing and quite high combustion temperatures. It did give more power, for a while. Then the heat build-up eventually took over. There may have been some detonation. On the dyno we would have seen the power 'fading' and would have had the option of making the mixture slightly richer (less power, but by cooling the piston we would be able to hold the power for longer) or to use a cooler-running plug or to retard the ignition slightly (in order to reduce the combustion temperature, although this would also have taken the edge off the performance.)

The original test had worked well, but we had finished before the end. We should have done several long, flat-out runs while keeping a check on engine temperature and plug condition. The rider would probably have noticed the drop in power, while the colour of the cylinder heads and piston crowns would have confirmed what was happening.

As the engine output increases, the chances of mechanical failure do too. Two-strokes are prone to several types of piston failure and in some conditions they will regularly destroy connecting rod bearings as well. Other problems are generally associated with the abnormally high speeds reached and vibration.

Piston crown Damage to the top of the piston is usually due to high temperatures – incorrect plug grade, ignition timing advanced, weak mixture, detonation – and will either take the form of a neat hole directly under the spark plug or the crown will melt and deform, usually in the region of the exhaust port. Severe detonation makes the piston look as if it has been peppered with small pieces of metal. Using one ring only, or having the top ring close to the crown of the piston will encourage the ring land to run hotter.

Piston skirt Scuffing or seizure here can be attributed to three basic causes, apart from running out of oil. First, metal-to-metal contact caused by the piston overheating or by the skirt clearance being too tight. Second, skirt varnishing caused by excessive piston clearance, or the wrong type of oil. Third, blow-by past the rings, due to ring failure or excessive clearance; the hot gas heats the piston skirt locally, causing it to deform and then to seize.

Ring failure Short ring life is normal but ring sticking can be caused by too

Fig. 77. Piston failure. Holed piston crown could be caused by pre-ignition, wrong plug grade, over-advanced timing or weak mixture

Fig. 78. Piston failure—seizure. Left, classic seizure, due to lubrication breakdown/overload. Right: seizure which was caused by piston failure when the top ring land melted. Such overheating could be caused by a rise in combustion temperature due to exhaust blockage/incorrect dimensions, weak mixture, pre-ignition or detonation

132

much groove clearance allowing deposits to build up, or by excessive piston rock damaging the ring lands or by the lands overheating/partially seizing. If the piston touches the cylinder head at high engine speed it may also damage the ring land, trapping the ring in its groove.

Small end failure Needle roller bearings will tolerate an amazing amount of abuse but overheating and lack of lubrication will eventually cause seizure. Often the piston pin seizes in the piston and may need careful lapping to get a reasonable fit.

Big end failure Seizure or a broken cage can either be caused by lack of lubrication or by the bearing breaking up under the stress of overspeeding. If this is the case then the speed will have to be limited or a stronger bearing used. It isn't unusual to go to a smaller diameter crank pin or hollow pins in order to reduce the inertia stress on the big end.

Stress raisers Connecting rods and crankshafts sometimes fail because manufacturing methods put notches or sudden changes of section in them and this increases the local stress distribution to beyond the metal's

Fig. 79. High speed strength. All potential stress raisers have been removed on these stock Suzuki crankshafts and the driven jackshaft. There are no sharp edges or corners, each change in section is carefully radiused and the root diameter of the gears and splines is not smaller than that of the shaft. Even the tapered shaft which carries the primary drive gear and its shock absorber is formed on a larger diameter than the main shaft. The big-end eyes of the connecting rods have large slots to allow oil mist to lubricate the roller bearings

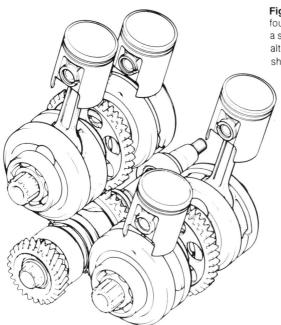

Fig. 80. Compact layout of the square four engine. Racers would normally use a straight-cut primary drive which, although noisier than the helical gear shown, absorbs less power

endurance limit. Then under cyclic stress the part suffers a fatigue failure, starting as a small crack at the point of maximum stress intensity – and the crack itself becomes a stress-raiser and rapidly increases in size. Removing all sharp-cornered notches and section changes is the only cure for this sort of failure – unfortunately it needs to be done before any heat treatment is carried out. Shot peening the surface of highly stressed components will increase their fatigue strength.

Crankshaft Two-strokes use pressed-together cranks, which can loosen or twist, particularly if they are not assembled correctly in the first place. A specialist crankshaft reconditioning firm will be able to build the crank within tolerances and may be able to increase the press fit of the crank pin. They should also be able to measure any twist or bending in the connecting rod. Even slight misalignment in the crank will cause rough running, vibration and will knock out the oil seals and main bearings.

Generator Most have a heavy flywheel keyed on to a shallow taper on the end of the crank and it is not unknown for the taper fit to loosen off under the effects of very high crank speeds. The generator should, if possible, be replaced by a smaller unit, or removed altogether. Failing that it should be accurately balanced and carefully lapped to the taper, using a fine abrasive paste, until it makes an accurate fit.

Clutch   The combination of high speed and extra load may prove to be too much for the standard clutch. The traditional way to increase pressure is to rebuild it with washers under the springs or an extra steel plate, but stronger springs are a better answer and will still leave full travel for the plates so that the clutch doesn't drag when the bike is at a standstill. Check that the slots in the clutch body have smooth edges and have not been damaged by the ears on the clutch plates.

Balance shafts   The thought of turning a balancer at imponderable revs persuades many tuners to remove them. However, this can cause severe vibration and will transmit the vibration to parts that were never meant to cope with it. When balance shafts are used, vibration is contained within the engine but the same forces still exist in the bearings holding the crankshaft and balance shaft – and in the portion of crankcase which contains the bearings. To avoid fatigue failure, any stress raisers in the crankcase should be removed (e.g. casting seams, sudden steps) and any webs etc. should have the largest possible radius. The surface could also be polished and bead blasted to remove any surface irregularities.

If a heavier piston is used then the balance shaft should be modified accordingly – usually by machining material from the unweighted side. This can be gauged by supporting the shaft in V-blocks, and making a support to hold the bob-weight horizontal on the weighing plate of a scale balance. The weight measurement is the out of balance force and if the piston weight is increased by x% then the out of balance force should also be increased by x%.

Exhausts   The front pipes or chamber sides often suffer from vibration and crack, especially if the engine is flexibly mounted while the chambers are bolted up solidly at the rear. Rubber mounting will help, as will exhaust bandage wrapped around the front pipe.

Appendix:

This appendix contains data and calculations which relate to the above topics, plus programs, some of which are written in BASIC for microcomputers and others which are suitable for programmable calculators such as the Texas Instruments TI53.

The computer programs are written in BBC BASIC or BASIC 2.

Road loads and gearing

The speed and torque produced at the crankshaft are altered by the gear train. If the speed is reduced by a factor of X then the torque is raised by a factor of X – the gear ratios simply give us several different values for X. The result is that the back wheel turns at a certain speed and generates a certain thrust at the tyre's contact patch. The speed/force here varies with (a) the engine's torque curve and (b) the gear ratio being used at the time. It is often useful to know how this will vary, if the power or engine speed is changed, or if the gearing is changed. It would help to know if the engine was becoming too peaky and needed close ratio gears; it would also help to know if it was correctly geared for top speed, and what sort of speed was attainable. This program does all of this, both in figures and graphically, and allows very rapid changes of gearing, or even of the tyre size, along with changes in the power output and engine speed range.

In the BASIC program, you need to feed in the bike's data – gearing, tyre size, etc. – and information on its power output, which can be taken from dyno tests, the manufacturer's claimed figures or magazine road tests. The graph displayed by the computer has an upwardly curving line which represents rolling resistance and air drag on the bike – making some assumptions about its size and shape. Lines 2130–2280 in the program give the drag force in relation to the bike's speed. The factor shown here actually tallies quite accurately with the measured performance of a Suzuki GSX-R1100; for bikes which are smaller/better streamlined, a lower factor can be used. Bigger machines, or those without streamlining, would need a larger factor. Where the road load curve crosses this line is the bike's maximum speed.

```
10 REM rl
20  REM  12/12/87:  17/4/89
25 REM Maximum speed 220 mi/h
30 REM BASIC2
40 REM jwr


100 GOSUB show: WINDOW #1 TITLE "RL.BAS"
101 REPEAT: rread=INKEY : UNTIL rread>-1
102 IF rread=114 THEN x=30: y=7: GOTO 140
103 CLS
105 INPUT AT(10;6)"Make/model: ",n$: GOSUB nam
106 PRINT: PRINT: PRINT AT(8) "This will be filed as "n1$". Is this OK? y/n";
107 REPEAT: nn=INKEY: UNTIL nn>-1: IF nn=110 THEN CLS: PRINT AT(10;2)"Full name: "n$:
INPUT AT(10;6) "File name (max. 8 characters, no space): ",n1$: IF n1${-3 TO}<>"RL1" THEN
n1$=n1$+".RL1"
110 PRINT: INPUT AT(10) "Number of gears: ",y
115 PRINT: PRINT AT(10) "For optimum speed, acceleration and gearshift"
116 PRINT AT (10) "predictions, give figures for the full engine speed"
117 PRINT AT (10) "range, up to maximum permissible speed.": PRINT
120 PRINT AT(10) "Number of power/speed entries:"
130 INPUT AT(10) "if less than 20, enter 0  ",x: IF x=0 THEN x=20


140  DIM n(x), hp(x), t(x), ld(x), v(y,x),  vr(y,x),  f(y,x),  g(y): aa=0
145  DIM fav(y+1,220), warn(220), a(220),  tim(220), thr(220),shift(y),  vsh(y)
146 IF rread=114 THEN GOSUB rea: aa=1
148 REPEAT
150 GOSUB menu
160 UNTIL d>=10
200 END


300 LABEL menu
305 WINDOW #1 TITLE "RL.BAS"
310 IF aa=0 THEN GOSUB inpg: GOSUB inpp
315 aa=aa+1
320 CLS
330 PRINT AT(12;4)"Change gearing data................1"
340 PRINT TAB(12) "Change power data..................2"
350 PRINT TAB(12) "Display  power/torque................3"
360 PRINT TAB(12) "Display thrust/road speed...........4"
365 PRINT TAB(12) "Display  acceleration *..............5"
368 PRINT TAB(12) "Display gearshift/traction data Ø...6"
370 PRINT TAB(12) "File data *........................7"
380 PRINT TAB(12) "Read data from file................8"
390 PRINT TAB(12) "Print data *.......................9"
400 PRINT TAB(12) "STOP...............................10"
405 PRINT: PRINT: PRINT TAB(10)"* use item 4 before selecting this"
406 PRINT TAB(10)"Ø use items 4 and 5 before selecting this"
407 PRINT: PRINT TAB(10)"Use screen 2 to see current specification."
410 INPUT d: ON d GOSUB inpg, inpp, pow, thr, acc,warn, fil, rea, pri
430 RETURN


500 LABEL show
```

```
505 OPTION DATE 1
510 CLS
520 WINDOW #1 FULL ON
540 WINDOW #1 OPEN
541 ELLIPSE 4150;3000,2500,0.7 WIDTH 5 COLOUR 2
542 PRINT AT(30;6) POINTS(20);"Road Loads"
543 PRINT AT(8;10) ADJUST (16);"Acceleration and terminal speed predictor"
545 PRINT AT(12;18) "Do you want to enter new data"
546 PRINT TAB(12)"or read data from file? n/r "
550 RETURN


600 LABEL inpg
610 CLS: WINDOW #1 TITLE "DRIVELINE SPECIFICATION"
615 IF aa>0 THEN 720
616 d1$=DATE$
620 INPUT AT (10;2) "Primary reduction: ",p
630 FOR i=1 TO y
640 PRINT AT (10) "Internal gear ratio "i;: INPUT ": ",g(i)
650 NEXT
660 INPUT AT (10) "Number of teeth on gearbox sprocket: ",t1
670 INPUT AT (10) "Number of teeth on wheel sprocket:    ",t2
680 PRINT "Do you know the rolling radius of the rear tyre? y/n "
690 REPEAT: e$=INKEY$: UNTIL e$>"":IF e$="n" THEN GOSUB tyre
700 PRINT:INPUT "Rolling radius of tyre (inches): ",r
710 IF aa=0 THEN 860
720 CLS: PRINT AT (10;4) "To change primary reduction press......1"
730 PRINT AT (21;5) "internal gears..............2"
740 PRINT AT (21;6) "gearbox sprocket............3"
750 PRINT AT (21;7) "wheel sprocket..............4"
760 PRINT AT (21;8) "tyre radius.................5"
775 PRINT AT (10;10)"No change............................6"
780 INPUT bb: ON bb GOTO 790, 800, 810, 820, 830, 870
790 PRINT AT (10;12) "Current primary reduction is "p;: INPUT "New ratio: ",p: GOTO 860
800 FOR i=1 TO y: PRINT AT (10) "Current gear "i" is "g(i);: INPUT "New ratio: ",g(i): NEXT:
GOTO 860
810 PRINT AT (10) "Current gearbox sprocket is "t1;: INPUT "New sprocket: ",t1: GOTO 860
820 PRINT AT (10) "Current wheel sprocket is "t2;: INPUT "New sprocket: ",t2: GOTO 860
830 PRINT AT (10) "Current rolling radius is "r: INPUT AT(10) "New radius: (enter 0 to see
tyre sizes) ",r: IF r=0 THEN GOSUB tyre: IF r=0 THEN 830 ELSE 860
860 PRINT: PRINT AT (8) "Any other changes? y/n ", :REPEAT: ee$=INKEY$: UNTIL ee$>"": IF
ee$="y" THEN 720
870 GOSUB spec: RETURN


900 LABEL inpp
910 CLS: WINDOW #1 TITLE "ENGINE OUTPUT"
920 PRINT AT (4;2) "Enter speed (rev/min) and engine output (bhp)."
980 PRINT "Enter speed; RETURN; output; RETURN. Enter 1 to stop input."
990 PRINT: PRINT AT (15) "Rev/min" AT (25) "bhp" AT (35)" torque, lb-ft"
1000 w=0
1010 REPEAT
1020 PRINT TAB(15) ">";:INPUT "", n(w);: IF n(w)=1 THEN 1030
1025 PRINT TAB(25)">";:INPUT"",hp(w);: t(w)=hp(w)*5252/(n(w)): PRINT TAB(35)
ROUND(t(w),1)
1030 w=w+1
1040 UNTIL n(w-1)=1 OR w=x OR hp(w-1)=1
1042 PRINT: PRINT "Are the figures OK? y/n"
1043 REPEAT: e$=INKEY$: UNTIL e$>"": IF LOWER$(e$)="n" THEN CLS: GOTO 980
```

```
1050 RETURN

1500 LABEL calc
1510 FOR j=1 TO y
1520    q=p*g(j)*t2/t1
1530 FOR i=0 TO w-2
1540    v(j,i)=0.00595*r*n(i)/q:   vr(j,i)=ROUND(v(j,i))
1550 f(j,i)=hp(i)*375/v(j,i):      REM vr() is rounded to whole mi/h
1560 NEXT:                         REM for use in calculating av force
1570 NEXT
1580 RETURN


1600 LABEL thr
1610 GOSUB calc
1620 CLS: WINDOW #1 TITLE "THRUST (lbf) v ROAD SPEED (mi/h)"
1630 PRINT AT(10;4) "Figures or graph? f/g ",: REPEAT: th$=INKEY$: UNTIL th$>"": IF aa<3
THEN 1640: IF th$="f" THEN 1640
1632 PRINT TAB(10) "Drag factor: option 1,2,3 or 4"
1634 INPUT AT(10) "or select 5 to see drag data: ",ad
1640 IF th$="g" THEN GOSUB gra: GOTO 1740
1650 CLS
1660 PRINT TAB(10) "Road speed" TAB(25)"Thrust"
1665 PRINT TAB(16) "mi/h" TAB(28) "lbf"
1670 PRINT
1680 FOR j=1 TO y
1690 FOR i=0 TO w-2
1700 PRINT USING "         ##.#";,v(j,i),f(j,i): GOSUB wait
1710 NEXT
1720 PRINT: PRINT TAB(8) USING "##.##&"; v(j,i-1)*1000/n(i-1) " mi/h per 1000
rev/min": PRINT: PRINT
1730 NEXT
1735 PRINT TAB(20)"Press any key to continue."
1736 IF INKEY$="" THEN 1736
1740 RETURN


1700 LABEL wait
1710 IF ww>0 THEN ww=ww+1: GOTO 1770
1720 yy=YPOS
1730 IF yy>750 THEN 1780
1740 PRINT AT(10;20) "Press SPACE bar to continue.";
1750 IF INKEY$<>" " THEN 1750
1760 ww=1: PRINT AT (10;20)"                       "
1770 IF ww=13 THEN ww=0: PRINT: GOTO 1740
1780 RETURN


1800 LABEL gra
1810 GOSUB drag
1815 WINDOW #1 TITLE "Thrust (lbf) v. road speed (mi/h)"
1820    sx=7000/v(y,w-2):  sx=sx/1.05
1830    i=0
1840    j=-1
1850 j=j+1: IF j=w-2 THEN 1880
1860 IF f(1,i)>=f(1,j) THEN 1850
1870 IF f(1,i)<f(1,j) THEN i=i+1: GOTO 1840
1880    sy=4000/f(1,i): i2=i  : sy=sy/1.05
1890 k=20: i=0: kk=ROUND(f(1,i)/50): kk=kk*10
1900 REPEAT: LINE (1000+k*i*sx);1000,(1000+k*i*sx);5000: i=i+1
```

```
1910  UNTIL  k*(i-1)*sx>=5500
1920  FOR j=0 TO i-1: MOVE  (800+k*j*sx);700: PRINT j*k: NEXT
1930  i=0
1940  REPEAT:  LINE  1000;(1000+kk*i*sy),8000;(1000+kk*i*sy):  i=i+1
1950  UNTIL  kk*sy*(i)>=4000
1960  FOR j=0 TO i-1: MOVE 400;(875+kk*j*sy): PRINT j*kk: NEXT
1970 FOR i=1 TO y
1980 FOR j=0 TO w-3
1990  LINE      1000+v(i,j)*sx;1000+f(i,j)*sy,1000+v(i,j+1)*sx;1000+f(i,j+1)*sy
2000 NEXT
2010 NEXT
2020  i=0:  REPEAT:  dr=a+b*i+c*i^2
2030  PLOT 1000+i*sx;1000+dr*sy MARKER 5 SIZE 1 COLOUR 1
2040  i=i+1: UNTIL  dr*sy>3800 OR  i>=v(y,w-2)
2045  MOVE 1200;1500: PRINT "Final drive: "t2"/"t1
2046  MOVE 1200;400: PRINT n$": "d1$
2050  MOVE 5500;400: PRINT "Press c to continue."
2060  IF INKEY$<>"c" THEN 2060
2070 RETURN


2100 LABEL drag
2102 IF ad>0 AND ad<4 THEN 2270
2105 WINDOW #1 TITLE "Drag factors."
2110 CLS
2120 PRINT AT(6;2) "Overall drag is assumed to take the form"
2130 PRINT AT(6) "   dr = a + bv + cv2"
2140 PRINT AT(6) " where a,b and c are constants accounting for rolling and"
2150 PRINT AT(6) "driveline drag, aerodynamic drag, inertia of the"
2160 PRINT AT(6) "machine and inertia of the rotating parts.""
2170 PRINT
2180 PRINT AT(6) "For large, unfaired bikes (Z1000J, GSXII00)"
2190 PRINT AT(6) "try values of a=16, b=0 and c=0.0105        = option 1"
2195 PRINT
2200 PRINT AT(6) "For large, faired bikes (GSX-R1100)"
2210 PRINT AT(6) "try values of a=16, b=0 and c=0.0091   = option 2"
2215 PRINT
2220 PRINT AT(6) "For small, faired bikes (TZR250)"
2230 PRINT AT(6) "try values of a=16, b=0 and c=0.008   = option 3"
2235 PRINT
2240 PRINT AT(6) "Or new values can be used.             = option 4"
2250 PRINT
2260 INPUT AT(10) "Option: ",ad
2270 IF ad=1 THEN a=16 : b=0: c=0.0105
2275 IF ad=2 THEN a=16 : b=0: c=0.0091
2280 IF ad=3 THEN a=16 : b=0: c=0.008
2290 IF ad=4 THEN INPUT AT(10)"Value for a: ",a;: INPUT AT(30)"Value for b: ",b;: INPUT
AT(50)"Value for c: ",c
2300 CLS: GOSUB spec
2310 RETURN


2400 LABEL acc
2405 IF f(1,1)=0 THEN GOSUB calc: GOSUB drag
2410 GOSUB inpw
2420 GOSUB avg
2430 CLS: t=0: i=0: WINDOW #1 TITLE "Speed (mi/h) v Time (s)"
2440 REPEAT
2480   a(i)=0.682*thr(i)/m
```

140

```
2490 tim(i)=1/a(i): t=t+tim(i): REM t() in seconds
2495 IF i=60 THEN t60=t
2500 i=i+1:UNTIL i-1=v(y,w-2) OR thr(i)<=0:vmax=i-1: etmax=t: j=i-1
2510 sx=6800/t: sy=4000/(vmax+10): i=0
2515 IF etmax>38 THEN n=4 ELSE n=2
2516 IF etmax>76 THEN n=8
2520 REPEAT
2530 LINE 1000+2*i*sx;1000,1000+2*i*sx;5000
2540 MOVE 800+n*i*sx;700: PRINT n*i
2550 i=i+1
2560 UNTIL n*i*sx>=7000
2570 i=0
2580 REPEAT
2590 LINE 1000;1000+i*sy,8000;1000+i*sy
2600 MOVE 450;900+i*sy: PRINT i
2610 i=i+20
2620 UNTIL i*sy>=4000
2625 dist=0: et=0: et4=0: d4=0: sp=0
2627 gr=sg
2630 FOR i=0 TO vmax-1
2640 dist=dist+(tim(i)*(2*i+1)/2)/3600: et=et+tim(i)
2650 LINE 1000+et*sx;1000+i*sy,1000+(et+tim(i+1))*sx; 1000+(i+1)*sy
2655 IF shift(gr)=i THEN 2656 ELSE 2660
2656 LINE 1000+et*sx;1000+i*sy,1050+et*sx;820+i*sy
2657 gr=gr+1
2660 IF dist>0.23 AND et4=0 THEN GOSUB quart: PRINT d4: PLOT 1000+et4*sx;1000+v4*sy
MARKER 2
2670 NEXT
2680 dmax=dist: IF dist>0.2494 THEN 2685
2681 REPEAT: dist=dist+(tim(i)*i)/3600: et=et+tim(i):UNTIL dist>0.2499
2682 et4=et: v4=i-1: REM for bikes which reach max speed inside 1/4mile
2684 dmax=ROUND(dmax,2):vmax=ROUND(vmax,1): etmax=ROUND(etmax,1):
et4=ROUND(et4,2):v4=ROUND(v4,1):t60=ROUND(t60,2)
2690 MOVE 4500;2000: PRINT "Max speed: "vmax"mi/h in "etmax"s "
2700 MOVE 6500;1750: PRINT "and "dmax" mi"
2710 MOVE 4500;1500: PRINT "SS 1/4 mi: "et4"s/"v4"mi/h"
2720 MOVE 4500;1250: PRINT "0-60mi/h:  "t60"s"
2725 MOVE 900;400: PRINT n$": "d1$
2730 MOVE 5800;400: PRINT "Press c to continue."
2740 IF INKEY$<>"c" THEN 2740
2750 RETURN

2900 LABEL inpw
2910 CLS: WINDOW #1 TITLE "DIMENSIONS AND LAUNCH DATA"
2912 IF wt=0 THEN 2920
2914 PRINT AT (10) "Is the data (weight, wheelbase, etc) the same as before? y/n: ",
2914 REPEAT: inp$=INKEY$: UNTIL inp$>"": IF inp$="y" THEN 3007
2920 INPUT AT(10) "Weight of bike and rider, lbf: ",wt:m=wt/32.2: PRINT
2930 PRINT TAB(10) "If the following data is not available,"
2940 PRINT TAB(10) "enter 0; the program will assume a typical value."
2950 PRINT
2960 INPUT AT(10) "Wheelbase, inches: ",wb: IF wb=0 THEN wb=56
2970 PRINT TAB(10) "Centre of gravity co-ordinates"
2980 INPUT AT(10) "Horizontal distance from rear wheel spindle, inches: ",cgx: IF cgx=0
THEN cgx=wb*0.5
2990 INPUT AT(10) "Height above ground, inches: ",cgy: IF cgy=0 THEN cgy=2*r+3: PRINT
3000 INPUT AT(10) "Coefficient of friction, tyre/road: ",mu: IF mu=0 THEN mu=1: PRINT
```

141

```
3002 PRINT AT(10)"(Note: to start in a gear other than first, alter line 3215.)":  INPUT
AT(10) "At what engine speed is the clutch engaged? ",cl: IF cl=0 THEN cl=n(0)
3003 IF cl>0 AND cl<n(0) THEN CLS: PRINT AT(10) "This is less than the first speed value.":
PRINT AT(10) "Enter 0 or a value greater than "n(0): GOTO 3002
3005 PRINT AT(10) "At what engine speed are gearshifts made? "
3006 INPUT AT(10) "If 0 is entered, the program will calculate optimum speeds. ",gs
3007 CLS: GOSUB spec
3010 RETURN


3100 LABEL avg
3105 FOR i=0 TO 220: warn(i)=0: thr(i)=0: FOR ii=1 TO y: fav(ii,i)=0: NEXT: NEXT
3110 i=1
3120 REPEAT: j=0
3130 REPEAT
3135    FOR v=vr(i,j) TO vr(i,j+1)
3140        fav(i,v)=f(i,j)+(f(i,j+1)-f(i,j))*(v-vr(i,j))/(vr(i,j+1)-vr(i,j))
3150 NEXT
3160    j=j+1
3170    UNTIL j-1=w-2
3200 i=i+1:IF i-1=y THEN 3210
3205 IF vr(i-1,j-1)<vr(i,0)-1 THEN GOSUB slip
3210 UNTIL i-1=y:
3215 sg=1: REM gear in which start is made
3218    speed=CEILING(cl*0.00595*r/(p*g(sg)*t2/t1))
3220 FOR v=0 TO speed
3225 fav(sg,v)=fav(sg,speed):  warn(v)=warn(v)+0.5
3226 IF fav(sg,v)>0.8*f(sg,i2) THEN fav(sg,v)=0.8*f(sg,i2):
3227 NEXT
3228 IF gs<>0 THEN GOSUB shift: GOTO 3340: REM condition for pre-chosen gearshift rpm.
3230 i=sg: v=0
3240 REPEAT
3250 REPEAT
3252    thr(v)=fav(i,v)
3255 GOSUB wheel
3260    thr(v)=thr(v)-(a+b*v+c*v^2)
3270    v=v+1
3275    UNTIL fav(i,v-1)<=fav(i+1,v-1) OR v-1=vr(i,w-2)
3280      shift(i)=v-1:  vsh(i)=shift(i)*p*g(i)*t2/(t1*0.00595*r)
3290    i=i+1
3295    UNTIL i-1=y OR thr(v-1)<=0
3340 RETURN


3400 LABEL shift
3410 i=sg:v=0
3420 FOR gear=i TO y-1: shift(gear)=ROUND(gs*0.00595*r/(p*g(gear)*t2/t1)):
vsh(gear)=gs: NEXT
3430 REPEAT
3440 REPEAT
3441 IF fav(i,v)=0 THEN fav(i,v)=f(i,0): IF FRAC( warn(v))<>0.5 THEN warn(v)=warn(v)+0.5
3442    thr(v)=fav(i,v)
3445 GOSUB wheel
3450    thr(v)=thr(v)-(a+b*v+c*v^2):  v=v+1
3460 UNTIL v-1=shift(i)
3470 i=i+1
3480 UNTIL i-1=y OR thr(v-1)<=0
3490 RETURN
```

142

```
3500 LABEL slip
3510 FOR v=vr(i-1,j-1) TO vr(i,0)-1
3520 fav(i,v)=f(i,0): warn(v)=warn(v)+0.5
3530 NEXT
3540 RETURN


3600 LABEL tyre
3610 WINDOW #2 FULL ON: CLS #2: WINDOW #2 TITLE "Tyre size and rolling radius": WINDOW
#2 OPEN
3620 SET ZONE 19
3625 PRINT #2 " Rolling radius in inches ±2%, based on ETRTO standard."
3630 PRINT #2: PRINT #2 " 15-inch","17-inch","18-inch","18-inch"
3640 PRINT #2
3650 PRINT #2 "140/90   11.96","2.50      10.73","3.60      11.62","100/80   11.66"
3660 PRINT #2 "150/90   12.30","2.75      11.11","4.10      12.11","110/80   11.96"
3670 PRINT #2 " ","3.00      11.38","4.25      12.91","120/80   12.26"
3680 PRINT #2 " 16-inch","4.50      12.59","4.25/85 12.46","130/80   12.57"
3690 PRINT #2 " "," "," "," "


3700 PRINT #2 "4.60      11.42","100/80   11.18","90/90     11.7","110/70   11.55"
3710 PRINT #2 "100/90   11.07","120/80   11.79","100/90   12.04","140/70   12.34"
3720 PRINT #2 "110/90   11.42"," ","110/90   12.38","150/70   12.61"
3730 PRINT #2 "120/90   11.75","110/90   11.90","120/90   12.72",""
3740 PRINT #2 "130/90   12.09","120/90   12.24","130/90   13.06"," "
3750 PRINT #2 "140/90   12.44","130/90   12.59","140/90   13.40","170/60   12.49"
3760 PRINT #2 "100/80   10.70"," "
3770 PRINT #2 "120/80   11.30","140/80   12.39"
3780 PRINT #2 "150/80   12.21"
3790 PRINT #2 TAB(40) "Press c to continue."
3800 IF INKEY$<>"c" THEN 3800
3810 WINDOW #1 OPEN
3820 RETURN


4000 LABEL pri
4050 GOSUB setup
4010 LPRINT n$, d1$: LPRINT
4020 LPRINT "Final drive "t2"/"t1: LPRINT
4030 LPRINT "speed, mi/h","thrust, lbf": LPRINT
4040 FOR j=1 TO y
4050 FOR i=0 TO w-2
4060 LPRINT ROUND(v(j,i),1),ROUND(f(j,i),1)
4070 NEXT: LPRINT "Gear "j": " ROUND(v(j,i-1)*1000/n(i-1),1)" mi/h per 1000rev/min":
LPRINT
4080 NEXT
4090 LPRINT: LPRINT
4100 RETURN


4850 LABEL pow
4860 pmax=hp(0):tmax=t(0):i=0
4870 REPEAT: i=i+1: pmax=MAX(pmax,hp(i)): UNTIL i=w-1: i=0
4875 REPEAT: i=i+1: tmax=MAX(tmax,t(i)): UNTIL i=w-1: mmax=MAX(pmax,tmax)
4880 sx=6500/n(w-2)
4890 sy=3800/mmax
4900 CLS: WINDOW #1 TITLE "Output (bhp and torque, lb ft) v. crank speed, rev/min x10^3"
4910 k=1000: i=0
4920 REPEAT: LINE 1000+k*i*sx;1000,1000+k*i*sx;5000: MOVE 800+k*i*sx;600: PRINT
i: i=i+1
```

143

```
4930 UNTIL k*(i-1)*sx >=6500
4940 i=0: IF sy<390 THEN k=10 ELSE k=1
4950 REPEAT: LINE 1000;1000+k*i*sy,8000;1000+k*i*sy: MOVE 400;1000+k*i*sy: PRINT
i*k: i=i+1: UNTIL k*i*sy>=3900
4960 FOR i=0 TO w-3: LINE
1000+n(i)*sx;1000+hp(i)*sy,1000+n(i+1)*sx;1000+hp(i+1)*sy:   NEXT
4970 FOR i=0 TO w-3: LINE 1000+n(i)*sx;1000+t(i)*sy,1000+n(i+1)*sx;1000+t(i+1)*sy:
NEXT
4980 MOVE 1000;300: PRINT n$": "d1$
4990 MOVE 5800;300: PRINT "Press c to continue."
5000 IF INKEY$ <>"c" THEN 5000
5010 RETURN


6000 LABEL quart
6010 et4=et: v4=i: d4=dist: counter=0: ii=i
6020 REPEAT
6030 v4=v4 + 0.1: counter=counter + 1
6040 et4=et4 + 0.1/a(ii)
6050 d4=d4+ v4/(a(ii)*36000)
6060 IF counter=9 THEN counter=-1 AND ii=ii+1
6070 UNTIL d4>0.2498
6080 RETURN


6100 LABEL nam
6110 n1$="": n2$=".RL1"
6120 le=LEN(n$)
6130 FOR i=1 TO le
6140 t$=n${i}
6150 IF ASC(t$)=32 THEN t$=""
6160 n1$=n1$+t$
6170 NEXT
6180 le=LEN(n1$)
6190 IF le>8 THEN n1$=n1${TO 3}+n1${-5 TO}
6195 n1$=n1$+n2$
6200 RETURN


6300 LABEL fil
6310 CLS: WINDOW #1 TITLE "Saving data on disc file"
6320 PRINT AT(10;3)"Put a formatted disc in drive B:"
6330 PRINT: PRINT AT(10)"The data will be stored in \RL\"n1$
6340 PRINT: PRINT AT(10)"To alter the filename, press A or press"
6342 PRINT AT(10) "any other key to continue."
6350 REPEAT:f1$=INKEY$: UNTIL f1$>"": IF LOWER$(f1$)="a" THEN GOSUB ns
6360 DRIVE "B"
6370 ON ERROR GOTO 6550
6380 CD \RL\
6400 ON ERROR GOTO 0
6420 IF n1$="" THEN n1$=n3$
6430 IF FIND$(n1$)>"" THEN REPEAT: GOSUB rname: UNTIL FIND$(n1$)="" OR r$="y"
6450 OPEN #3 OUTPUT n1$
6460 PRINT #3,n$
6462 PRINT #3,d1$
6470 PRINT #3,w,y,p,t1,t2,r,wt,wb,cgx,cgy,mu,cl,gs
6480 FOR i=1 TO y
6490 PRINT #3,g(i)
6500 NEXT
6510 FOR i=0 TO w-2
```

144

```
6520  PRINT  #3,n(i),hp(i),t(i)
6530 NEXT
6540 CLOSE #3: GOTO 6570
6550 IF ERR=133 THEN GOSUB drctry
6560 RESUME NEXT
6570 RETURN

6600 LABEL rname
6610 PRINT
6620 PRINT AT(10)"A file "n1$" already exists."
6630 INPUT AT(10)"Do you want this file to replace it? y/n ",r$
6640 IF r$="y" THEN 6670
6650  mm=ASC(n1${-1})+1
6660  n1${-1}=CHR$(mm)
6670 RETURN

6700 LABEL drctry
6710  MD \rl\
6720  CD \rl\
6730 RETURN

6800 LABEL rea
6810 CLS: WINDOW #1 TITLE "READ DATA FROM FILE"
6820 DRIVE "B"
6830 PRINT AT(10;6)"Put the file disc into drive B:"
6840 PRINT AT(10)"Press a key when ready."
6850 IF INKEY$="" THEN 6850
6855 ON ERROR GOTO 6990
6860  CD \rl\
6870 ON ERROR GOTO 0
6880 PRINT: PRINT AT(10)"Note that file names are compressed into a maximum": PRINT
AT(10)"of eight characters, with no spaces, and have ": PRINT AT(10)"an extension
.RLx":GOSUB look
6890 PRINT: REPEAT: INPUT "Filename for machine, or Q to quit: ",n3$
6895 IF n3$="q" THEN 6905
6900 IF FIND$(n3$)="" THEN PRINT "No file for "n3$
6905 UNTIL FIND$(n3$)>"" OR n3$="q": IF n3$="q" THEN 7000
6910 PRINT: PRINT "Reading "n3$
6920 OPEN #3 INPUT n3$
6930 INPUT #3,n$
6932 INPUT #3,d1$
6940   INPUT   #3,w,y,p,t1,t2,r,wt,wb,cgx,cgy,mu,cl,gs
6950 FOR i=1 TO y
6960  INPUT #3,g(i)
6965 NEXT
6970 FOR i=0 TO w-2
6975  INPUT  #3,n(i),hp(i),t(i)
6980 NEXT
6985 CLOSE #3
6986  m=wt/32.2
6987 CLS:PRINT: PRINT AT(10)n$
6988 tim=TIME:REPEAT:UNTIL TIME>tim+500:GOTO 7020
6990 IF ERR=133 THEN PRINT AT(10)"There is no \RL\ directory on this disc."
7000 PRINT TAB(10)"Try another disc or quit? a/q ";: INPUT re$
7010 IF re$="a" THEN 6810
7020 RETURN
```

```
7100 LABEL ns
7110 PRINT "Existing filename: "n1$
7120 INPUT "New filename, including extension: ",n1$
7130 PRINT "Existing full name: "n$
7140 INPUT "New name: ",n$
7150 RETURN


7200 LABEL wheel
7210 IF thr(v)*cgy>wt*cgx THEN warn(v)= warn(v)+1: thr(v)=wt*cgx/cgy
7220 IF thr(v)>wt*mu THEN warn(v)=warn(v)+2: thr(v)=wt*mu
7230 RETURN


7400 LABEL warn
7405 WINDOW #1 TITLE "Gearshift and traction data"
7410 CLS
7415  sxx=6300/vmax
7420 FOR i=0 TO vmax
7430 IF INT(warn(i))=1 OR INT(warn(i))=3 THEN
LINE(1500+i*sxx);1000,(1500+(i+1)*sxx);1000 WIDTH 5 COLOUR 1
7440 IF INT(warn(i))=2 OR INT(warn(i))=3 THEN
LINE(1500+i*sxx);1500,(1500+(i+1)*sxx);1500 WIDTH 5 COLOUR 2
7450 IF FRAC(warn(i))=0.5 THEN LINE(1500+i*sxx);2000,(1500+(i+1)*sxx);2000 WIDTH 5
COLOUR 11
7460 NEXT
7470 LOCATE 1;17: PRINT "Wheelie"
7480 LOCATE 1;15: PRINT "Wheelspin"
7490 LOCATE 1;13: PRINT "Clutch slip"
7500 FOR i=0 TO vmax STEP 10
7510  LINE(1500+i*sxx);1000,(1500+i*sxx);2100
7520 MOVE (1300+i*sxx);700: IF i<vmax-15 THEN  PRINT i
7530 NEXT
7540 LOCATE 10;1: PRINT "Gearshift     ","rev/min","mi/h"
7545 IF gs=0 THEN PRINT AT(10) "(calculated)"
7550 PRINT
7560 FOR i=1 TO y-1
7570 PRINT AT(10) i" to "i+1,ROUND(vsh(i)),shift(i)
7580 NEXT
7580 PRINT AT(12;19)"Road speed, mi/h" AT(50;19) "Press any key to continue"
7590 IF INKEY$="" THEN 7590
 7600 RETURN


8000 LABEL look
8010 PRINT "Here are the files on this disc:"
8020 FILES
8030 RETURN


8100 LABEL setup
8110 LPRINT CHR$(15)
8120 LPRINT CHR$(27)+"I"+CHR$(15)
8130 LPRINT CHR$(27)+"N"+CHR$(5)
8140 RETURN


8200 LABEL spec
8202 WINDOW #2 SIZE 27,30: WINDOW #2 PLACE 400,0
8203 WINDOW #2 TITLE "CURRENT SPECIFICATION"
8210 CLS #2: PRINT #2
8220 PRINT #2 " Final drive     "t2"/"t1
```

```
8230 PRINT #2 " Primary        "ROUND(p,3)
8240 PRINT #2 " Roll radius    "r
8250 PRINT #2
8260 PRINT #2 " Wheelbase      "wb
8270 PRINT #2 " cgx - r wheel  "cgx
8280 PRINT #2 " cgy            "cgy
8290 PRINT #2 " Total weight   "wt
8300 PRINT #2 " μ              "mu
8310 PRINT #2
8320 PRINT #2 " Start gear     "sg
8330 PRINT #2 "Clutch dump rpm "cl
8340 PRINT #2
8350 PRINT #2 " Total drag"
8360 PRINT #2 "  =a + bv + cv2"
8370 PRINT #2 "     a="a
8380 PRINT #2 "     b="b
8390 PRINT #2 "     c="c
8400 RETURN
```

Fig. 81. Road load curves for a Suzuki moto-crosser. The thrust available at the rear wheel is shown for stock gearing, which would only allow the machine to reach a speed of 65 mph in top gear. Raising the gearing by changing the gearbox and rear wheel sprockets as shown, would produce the top-gear thrust shown by the dotted line – and a top speed of about 94 mph. The force due to air drag would obviously change with the frontal area of the bike and any streamlining which was used

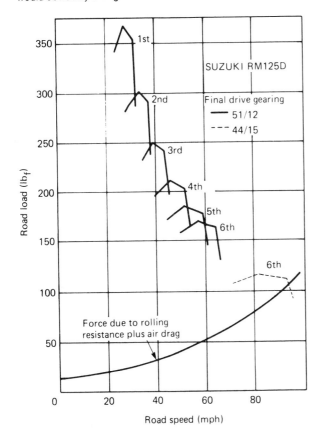

While the computer does it almost instantaneously, the same results can be achieved using an ordinary calculator. The calculation is:

$$F = bhp \times 375/v$$

where: F — thrust at wheel, in lbf
bhp — engine horsepower

$$v = KNG/W$$

where: v — road speed, in mi/h
N — crank speed, in rev/min
G — number of teeth on gearbox sprocket
W — number of teeth on rear wheel sprocket

$$K = 0.00595R/PG_1$$

where: P — primary reduction ratio
G_1 — internal gear ratio
R — rear tyre's rolling radius, in inches

F and v can be calculated for several power and rpm positions, for each gear, and plotted on a graph, with F using the y-axis. The TI53 program, after entering the programming mode, is:

First calculate K and G/W to four decimal places.

Step	Key	Symbol	
00	55	x	(input N rev/min)
01	83	.	
02	( )	K1	(insert K to four decimal places)
03	( )	K2	
04	( )	K3	
05	( )	K4	
06	55	x	
07	83		
08	( )	G/W	(insert G/W to four decimal places)
09	( )	G/W	
10	( )	G/W	
11	( )	G/W	
12	85	=	
13	41	STO	(displays v in mi/h)
14	81	R/S	(input bhp)
15	55	x	
16	74	3	
17	52	7	
18	63	5	
19	45	÷	
20	51	RCL	
21	85	=	(displays F in lbf)
22	81	R/S	
23	31	2nd	
24	81	RST	

(Followed by LRN, 2nd, RST to leave programming mode and reset.)

148

Compression ratio

To calculate the compression ratio C:

$$C = (V_c + V_{sw})/V_c$$

$$C_{new} = (V_c \pm Ah + V_{sw})/(V_c \pm Ah)$$

where: V_c – combustion chamber volume

V_{sw} – volume displaced by piston, *either* from BDC *or* from exhaust closing

h   – depth of material removed (or added) at gasket face

A   – area of cylinder bore $(3.14 \times (bore)^2/4)$

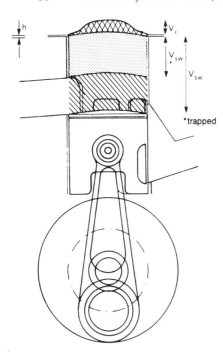

Fig. 82. Two-stroke engine proportions. The compression ratio is calculated from the volumes indicated: V_{sw} – swept volume (or trapped volume after exhaust has closed); V_c – combustion chamber volume. The height h is the amount varied by machining the head/barrel gasket surface or by changing the head gasket

The TI 53 program for calculating the change in compression ratio is:

Step	Key	Symbol	
00	55	x	(input h)
01	( )	A	
02	( )	A	
03	( )	A	
04	( )	A	
05	65	–	
06	( )	V_c	
07	( )	V_c	
08	( )	V_c	
09	( )	V_c	
10	85	=	
11	41	STO	
12	65	–	

Step	Key	Symbol
13	( )	V_{sw}
14	( )	V_{sw}
15	( )	V_{sw}
16	( )	V_{sw}
17	( )	V_{sw}
18	85	=
19	45	÷
20	51	RCL
21	85	=
22	81	R/S
23	31	2nd
24	81	RST

Input any value of h and the program will display the new compression ratio. It assumes h is being removed from the gasket face; if it is being added, e.g. extra gasket, then input −h.

Piston travel v. crank rotation

It is often useful to be able to convert crankshaft degrees into piston height before TDC. This calculation does it and the following TI 53 program will display piston height for any angle which is fed into the calculator. From this a graph of piston height v crank angle can be drawn, allowing instant conversion from degrees to mm and vice versa.

If the piston height before/after TDC is d then:

$$d = \frac{S}{2} + L - \frac{S}{2} \cos X - L \sin\left[\cos^{-1}\left(\frac{S}{2L} \sin X\right)\right]$$

where: S − stroke (mm)
   L − rod length (mm)
   X − crank angle B/A TDC (deg)

The following program will run on a TI 53 calculator; first:

(i) calculate $\frac{S}{2L}$ (= A) to 3 digits (including decimal point)

(ii) calculate $\frac{S}{2}$ (= B) to 4 digits

(iii) calculate $\frac{S}{2} + L$ (= C) to 5 digits

(Note that L is usually 2S).

If the calculator will accept a program of more than 31 steps, then more digits may be used for A, B or C. The final display (d) will be shown as a negative number.

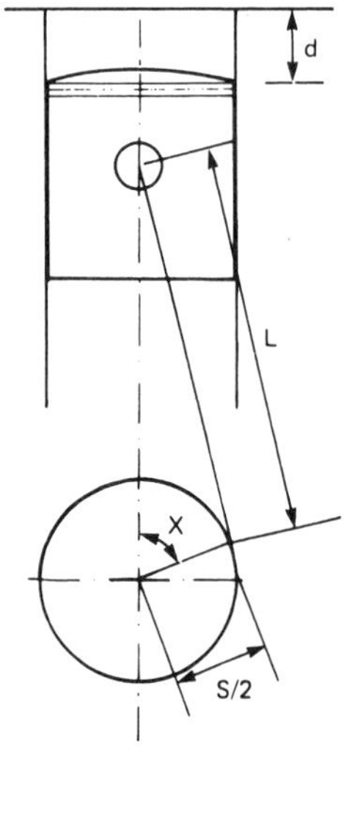

Fig. 83. Piston travel. The distance before/after TDC (d) is related to the angle (X) through which the crank has turned. S represents the stroke and L the length between centres of the connecting rod

150

Step	Key	Function	
00	41	STO	input x (degrees B/A TDC)
01	22	sin	
02	55	x	
03	83	.	(decimal point)
04	( )	A	enter A (usually 0.25)
05	( )	A	
06	85	=	
07	31	2nd	
08	23	cos^{-1}	
09	22	sin	
10	55	x	
11	( )	L	enter L
12	( )	L	
13	( )	L	
14	75	+	
15	( )	B	enter B
16	( )	B	
17	( )	B	
18	( )	B	
19	55	x	
20	51	RCL	
21	23	cos	
22	65	−	
23	( )	C	enter C
24	( )	C	
25	( )	C	
26	( )	C	
27	( )	C	
28	85	=	display d
29	81	R/S	
30	31	2nd	
31	81	RST	

Fig. 84. Graph of piston travel produced by the listed program. The effect of connecting rod angularity means that the piston spends longer at BDC than TDC (and that its acceleration is less in the BDC region)

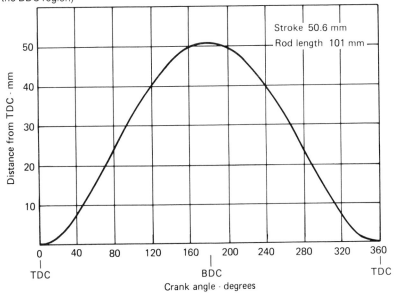

```
10 REM PT
20 REM  24/8/85
30 REM BBC B
40 REM PISTON HEIGHT v CRANK ROTATION
45 REM JWR
50 CLS
60 PRINT: PRINT "Enter data in mm and"
65 PRINT "crankshaft degrees, B/ATDC"
70 INPUT "Stroke: "S
80 PRINT "Rod length: "
90 INPUT "If not known, enter 0: "L: IF L=0 THEN L=2*S
100 INPUT "Maximum crank angle: "X1
110 INPUT "Minimum crank angle: "X2
120 INPUT "Incremental step: "X3
130 CLS
135 VDU 14: REM prevents screen scrolling until SHIFT is pressed
140 @%=&2020B: REM print format: 2 decimal places, column width 11 (hex B)
characters
145 ON ERROR GOTO 370
150 A=S/2+L
160 Z1=RAD(X1): Z2=RAD(X2): Z3=RAD(X3)
170 PRINT SPC(6) "Angle",SPC(3) "Distance"
180 PRINT SPC(3) "from TDC", SPC(2) "below TDC"
190 FOR Z=Z1 TO Z2 STEP Z3
200   P=S/2*COS(Z)
210   R=S/2*SIN(Z)
220   Q=SQR(L^2-R^2)
230   D=A-P-Q
249   PRINT DEG(Z), D
250 NEXT
260 PRINT TAB(23) "Press any key"
265 PRINT TAB(25) "to continue"
270 IF GET>0 THEN 280
280 CLS
285 PRINT:PRINT:PRINT
290 PRINT "To alter the STROKE press............1"
300 PRINT: PRINT SPC(13) "ROD LENGTH.....2"
310 PRINT: PRINT SPC(13) "CRANK ANGLE...3"
320 PRINT: PRINT SPC(10) "To STOP ..........4"
330 INPUT M: CLS
340 IF M=1 THEN PRINT "Stroke:" S "mm": INPUT "New stroke:" S: GOTO 150
350 IF M=2 THEN PRINT "Length:" L "mm": INPUT "New length:" L: GOTO 150
360 IF M=3 THEN 100
365 IF M>3 THEN 380
370 IF ERR=21 THEN PRINT "The rod must be longer": PRINT "than the crank
throw."
380 END
```

Piston velocity and acceleration

The piston's mean velocity is v, where

$$v = 2sN$$

where s – stroke
      N – crank speed

152

if s is in metres and N is in rev/s, then the mean velocity will be in m/s. The generally accepted maximum for mechanical reliability is around 20 m/s.

The force on the piston is the product of its mass and its acceleration and it can be important to know how much this has changed when either the piston mass or the engine speed range have been altered.

The acceleration of the piston is a, where

$$a = b^2 r \left(\cos X + \frac{r}{L} \cos 2X \right)$$

Where: b — crank velocity in radians per second
  r — stroke/2
  L — rod length between centres
  X — angle made by crank to the line of the stroke.

when X = 0 (i.e. at TDC), a will be maximum

$$a_{max} = b^2 r \left(1 + \frac{r}{L} \right)$$

(Note, b = N × 0.1047, where N is the engine speed in rev/min)

If the piston mass is raised, then to maintain the same maximum stress, the safe engine speed will have to be lowered. If the original redline was at N_1 rev/min, then the new one will be N_2 where

$$N_2 = N_1 \sqrt{M_1/M_2}$$

where: M_1 — old mass (piston, pin, rings, top ⅔ of rod)
  M_2 — new mass.

Port time-area

A graph of the piston motion will show how it would uncover a port such as the exhaust or one of the scavenge ports. The window would only be fully open for a small portion of the timing period; during the rest of the time, the open size of the port would be continually changing. Both the size of the port *and* the time it spends open determine how much gas will flow through it (assuming that other factors, like the pressure, the discharge coefficient and so on, remain the same).

This gives rise to the notion of time-area, that is, the total area exposed multiplied by the time which it spends open. For example, a port measuring 2 cm^2 which was open for 3 seconds would have a time-area of 6 seconds-cm². In terms of its flowing capacity it would be the same as a port of 3 cm^2 which was open for 2 seconds.

The difficulty, in real engines, is in calculating the actual area of port which is open – but it is just the sort of calculation which computers are good at. The BASIC program which follows can calculate time-area integrals rapidly and accurately. It also calculates piston displacement, piston area and the port width as a percentage of the bore and as an arc of the bore (70 per cent, or 90 degrees, is considered to be the maximum for ring reliability).

The time-area is displayed in units of seconds-mm^2 and, in order to compare engines of different sizes, the specific time area in s-mm^2/cm^3 (that is, the time-area divided by the displacement).

This means that an engine being evaluated for tuning can be compared to state-of-the art racers, to its competitors or to a competition variant built by the same manufacturer, and the program automatically allows comparisons between engines with different bore/stroke ratios, as well as capacities.

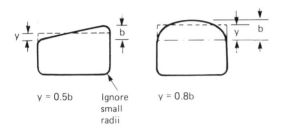

y = 0.5b   Ignore small radii   y = 0.8b

Fig. 85. Equivalent port shapes for use with the time-area program. Where the true dimensions are shown by *a,b* an approximately equivalent rectangular shape is shown by *x,y.* Bottom right: Calculate the time-area for *x,y (= TA)* and then calculate the time-area for *a,b (= ta).* The total time-area for this shape of port is then *TA − 2ta*

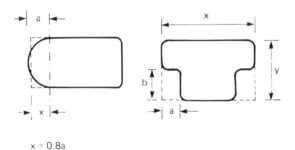

x = 0.8a

The program gives a clear indication of the state of tune of the engine, and even lets it be compared to itself – with modified porting or with a larger bore size. Engine data can be changed rapidly, to see the effects of changing the port width, height, engine speed, rod length or any of the basic engine dimensions.

Note: for bridged or multiple ports, do not enter a width value which is greater than the bore; use one half and then double the answer. The program assumes the ports to be rectangular, if they are not, then it will be necessary to devise an equivalent rectangular shape, as shown in the diagrams.

To run the program, enter the data requested. If the rod length is not known, the program will assume it to be twice the stroke. The port timing may be given in one of three ways – as height above BDC (port height) in mm, as the opening point in mm after TDC or in crankshaft degrees after TDC. It is only necessary to give one of these, entering zero for the other two.

The time-area can then be varied with engine speed, or one speed can be selected and the time-area varied as a function of port width, height, stroke, etc. There are several ways of using the results. One is to take the time-area at peak torque and use the computer to show what size and timing of the ports would be necessary to maintain this time-area at a higher speed. Another way is to compare the engine with a more highly tuned machine and match its time-area. Obviously the carburettor, exhaust, etc. would also have to be uprated to match the new porting.

```
10 REM TA2
20   REM   22/11/87:20/1/88
30 REM BASIC2
40 REM JWR

45 ON ERROR GOTO er
50 GOSUB show
60 GOSUB inpa
65 GOSUB inpb
80 GOSUB vary
90 END

100 LABEL show
110 CLS #1: CLS #2
160 WINDOW #1 FULL ON
170 WINDOW #2 SIZE 27,30
180 WINDOW #2 PLACE 440,0
190 WINDOW #1 OPEN: WINDOW #2 OPEN
200 SET ZONE 15
210 RETURN

300 LABEL inpa
310 PRINT:PRINT
320 PRINT "Enter all dimensions in"
330 PRINT "in mm, degrees or rev/min."
340 PRINT: PRINT "If the data is not known, enter 0."
350 PRINT: INPUT "Bore:          ",b
360 INPUT "Stroke:        ",s: IF s=0 OR b=0 THEN PRINT "Insufficient data.":PRINT:GOTO 350
370 INPUT "Rod length:     ",l: IF l=0 THEN l=2*s
380 IF p>0 OR po>0 OR h>0 THEN GOSUB spec
390 RETURN

500 LABEL inpb
510 PRINT: INPUT "Port opens, deg ATDC: ",p: pd=p:p=RAD(p)
520 IF p=0 THEN INPUT "Port opens, mm ATDC: ",po
530 IF po=0 AND p=0 THEN INPUT "Port height:       ",h
540 IF p=0 AND po=0 AND h=0 THEN PRINT "Insufficient data.":PRINT:GOTO 510
550 INPUT "Port width:           ",w: IF w=0 THEN w=0.7*b:PRINT "A port width equal to
0.7*bore is " USING "##.##&"; w"mm."
560 GOSUB spec
570 PRINT: PRINT: PRINT "Press any key"
575 IF INKEY$ ="" THEN 575
580 RETURN

600 LABEL spec
605 CLS #2: PRINT #2 "Current specification:": PRINT #2: GOSUB calc
```

```
610 PRINT #2 "Bore        "b"mm"
620 PRINT #2 "Stroke      "s"mm"
630 PRINT #2 "Rod length "l"mm"
640 PRINT #2 "Piston area " USING "##.##&"; pa" mm2"
650 PRINT #2 "Piston dispt " USING"##.##&"; d"cc"
660 PRINT #2 "Port opens  " USING "##.##&"; pd"deg ATDC"
670 PRINT #2 "Port opens  " USING "##.##&"; po"mm ATDC"
680 PRINT #2 "Port height " USING "##.##&"; h"mm"
690 PRINT #2 "Port width " USING "##.##&"; w"mm"
700 PRINT #2 " corresponds to " USING "##.##&"; r"%"
710 PRINT #2 "of bore and subtends"
720 PRINT #2 "an angle of " USING "##.##&"; q"deg."
730 RETURN


800 LABEL calc
810  y=0
820  pa=PI*b^2/4
830  d=pa*s/1000
840 IF p>0 THEN 900
850 IF p=0 AND po=0 THEN 880
860 IF p=0 THEN i=0: REPEAT: i=i+0.01: UNTIL s/2*(COS(i))+SQR(l^2-
(s/2*SIN(i))^2)<=s/2+l-po
870 pd=DEG(i-0.01): h=s-po: GOTO 930
880 i=0:REPEAT: i=i+0.1:  UNTIL s/2*COS(i)+SQR(l^2-(s/2*SIN(i))^2)<=l-s/2+h
890 pd=DEG(i-0.01): po=s-h:GOTO 930
900 IF h=0 THEN y=COS(p)*s/2+SQR(l^2-(s/2*SIN(p))^2): pd=DEG(p)
910 IF h=0 THEN h=s/2-l+y: IF po=0 THEN po=s-h
930 q=DEG(2* ASIN(w/b))
940  r=w/b*100
950 RETURN


1000 LABEL vary
1010 CLS #1
1020 PRINT AT(10;4) "Which item do you want to vary?"
1030 PRINT AT(16;8) "Crank speed....1"
1040 PRINT AT(16;9) "Port width.....2"
1050 PRINT TAB(16)  "Port height....3"
1060 PRINT TAB(16)  "Rod length.....4"
1070 PRINT TAB(16)  "Stroke.........5"
1080 PRINT TAB(16)  "STOP...........6"
1090 LOCATE 10;17: INPUT "Item: ",x
1100 ON x GOSUB spe,wid,hei,rod,str
1110 IF x <>6 THEN 1010
1120 RETURN


1200 LABEL spe
1210 CLS #1:
1220 LOCATE 10;2:INPUT "Minimum speed: ",f
1230 LOCATE 10;3:INPUT "Maximum speed: ",g
1240 LOCATE 10;4:INPUT "Increment:     ",j
1250 PRINT:PRINT
1260 PRINT TAB(4)"rev/min" TAB(17)"s-sq mm" TAB(30) "s-sq mm/cc"
1270 PRINT TAB(34) "x10^-3": PRINT
1280 FOR n=f TO g STEP j
1290  n=n*PI/30
1300 GOSUB ta
1310  n=n*30/PI
```

156

```
1320 PRINT TAB(4) n, USING "##.##        "; ta,ta/d*1000
1325 GOSUB wait
1330 NEXT
1340 GOSUB cha
1350 RETURN

1500 LABEL ta
1510  ta=0
1520  i=0.1       :REM this may be varied
1530  t=i/n
1540 FOR e=0 TO 2*PI STEP i
1550  c=s/2*COS(e)+SQR(l^2-(s/2*SIN(e))^2)
1560  a=s/2-l+c
1570 IF a>h THEN 1590
1580  ta=ta+t*w*(h-a)
1590 NEXT
1600 RETURN

1700 LABEL wid
1710 CLS #1: w1=w
1720 LOCATE 10;2: INPUT "Crank speed: ",n: n=n*PI/30
1730 LOCATE 10;3: INPUT "Minimum width: ",f
1740 LOCATE 10;4: INPUT "Maximum width: ",g
1750 LOCATE 10;5: INPUT "Increment: ",j
1760 PRINT
1770 PRINT TAB(4) "Width, mm" TAB(17) "s-sq mm" TAB(30) "s-sq mm/cc"
1780 PRINT TAB(34) "x10^-3"
1790 PRINT
1800 FOR w=f TO g STEP j
1810 GOSUB ta
1820 PRINT TAB(4) w, USING "##.##         ";ta,ta/d*1000
1825 GOSUB wait
1830 NEXT
1835  w=w1
1840 GOSUB cha
1850 RETURN

1900 LABEL cha
1910 PRINT:PRINT
1920 PRINT "Press c to continue,"
1930 PRINT "m to change engine spec"
1940 INPUT "or p to alter port details. ",a$
1950 IF a$="m" THEN GOSUB inpa
1960 IF a$="p" THEN p=0:po=0: h=0: GOSUB inpb
1970 RETURN

2000 LABEL hei
2010 CLS #1: h=h1
2020 LOCATE 10;2: INPUT "Crank speed: ",n:n=n*PI/30
2030 LOCATE 10;3: INPUT "Minimum height: ",f
2040 LOCATE 10;4: INPUT "Maximum height: ",g
2050 LOCATE 10;5: INPUT "Increment: ",j
2060 PRINT
2070 PRINT TAB(4) "Height, mm" TAB(17) "s-sq mm" TAB(30) "s-sq mm/cc"
2080 PRINT TAB(34) "x10^-3"
2090 FOR h=f TO g STEP j
2100 GOSUB ta
```

```
2110 PRINT TAB(4) h, USING "##.##          ";ta,ta/d*1000
2115 GOSUB wait
2120 NEXT
2130 h=h1
2140 GOSUB cha
2150 RETURN


2200 LABEL rod
2205 CLS #1
2210 PRINT "The program assumes that the port is "
2220 PRINT "adjusted to match the new deck height"
2230 PRINT "when the rod length is changed. The"
2240 PRINT "height of the port is considered constant.": PRINT
2250 LOCATE 10: INPUT "Crank speed: ",n:n=n*PI/30
2260 LOCATE 10: INPUT "Minimum rod length: ",f
2270 LOCATE 10: INPUT "Maximum rod length: ",g
2280 LOCATE 10: INPUT "Increment: ",j
2290 PRINT
2300 PRINT TAB(4) "Length, mm" TAB(20) "s-sq mm" TAB(33) "s-sq mm/cc"
2310 PRINT TAB(37) "x10^-3"
2320 PRINT
2330 l1=l
2340 FOR l=f TO g STEP j
2350 GOSUB ta
2360 PRINT TAB(4) l, USING "##.##          ";ta,ta/d*1000
2365 GOSUB wait
2370 NEXT
2380 l=l1
2390 GOSUB cha
2400 RETURN


2500 LABEL str
2505 CLS #1
2510 PRINT "The program assumes that the port is "
2520 PRINT "adjusted to match the new deck height"
2530 PRINT "when the stroke is changed. The"
2540 PRINT "height of the port is considered constant.": PRINT
2560 LOCATE 4: INPUT "Crank speed: ",n:n=n*PI/30
2570 LOCATE 4: INPUT "Minimum stroke: ",f
2580 LOCATE 4: INPUT "Maximum stroke: ",g
2590 LOCATE 4: INPUT "Increment: ",j
2600 PRINT
2610 PRINT TAB(4) "Stroke, mm" TAB(17) "s-sq mm" TAB(30) "s-sq mm/cc"
2620 PRINT TAB(34) "x10^-3"
2630 s1=s
2640 FOR s=f TO g STEP j
2650 GOSUB ta
2660 PRINT TAB(4) s, USING "##.##          ";ta,ta/d*1000
2665 GOSUB wait
2670 NEXT
2680 s=s1
2690 GOSUB cha
2700 RETURN


2800 LABEL wait
2810 IF aa>0 THEN aa=aa+1: GOTO 2870
2820 yy=YPOS
```

```
2830 IF yy>750 THEN 2880
2840 PRINT AT (10;20) "Press SPACE bar to continue.";
2850 IF INKEY$ <>" " THEN 2850
2860 aa=1: PRINT AT (10;20)"                    "
2870 IF aa=13 THEN aa=0: PRINT: PRINT: GOTO 2840
2880 RETURN

LABEL er
IF ERR=106 THEN PRINT "Check that the port width is not greater": PRINT "than the bore (if
necessary use half": PRINT " the width and then double the result)": PRINT "and check that the
rod length is greater": PRINT "than the crank throw."
IF ERR<>106 THEN PRINT "Error number " ERR
 PRINT "Press a key to restart": REPEAT:UNTIL INKEY$ > ""
GOTO 45
```

Approximation

If you don't have access to a computer, a close approximation to the time-area values for a rectangular port is

time-area = KHW/N s-mm^2

where: H is the port height in mm; W is the port width in mm; N is the engine speed in rev/min; and K is proportional to the port's duration

duration	130°	140°	150°	160°	170°	180°	190°	200°	210°
K	14.35	15.40	16.45	17.50	18.65	19.65	20.70	21.75	22.80

The program deals with ports which are open around BDC. Another program, with the title TA2A, calculates the time-area for a piston-controlled intaked, which opens around TDC and which can be altered by changing the size or shape of the port and by altering the length of the piston skirt. This only applies to piston-controlled intakes; where a reed valve is used to control the port, it can be open for a full 360°, the reed opening and closing on demand. It is impossible to improve on this time-area, at least on the time aspect of it.

```
10 REM TA2A
20 REM 29/10/86, revised 29/12/86
30 REM BASIC2 2/5/88: 8/5/88
40 REM JWR
50 REM Intake time-area

100 CLS: m=0
105 OPTION DEGREES
107 WINDOW #1 FULL: WINDOW OPEN: WINDOW TITLE "TA2A.BAS"
110 GOSUB inpa
120 GOSUB inpb
130 REPEAT
140 GOSUB menu
150 UNTIL m=8
160 END

200 LABEL inpa
205 CLS
210 PRINT AT(10;2) "Enter data in mm and crank degrees relative to TDC."
```

```
220 PRINT: PRINT
230 INPUT AT (10)"Bore: ",b
240 INPUT AT(10)"Stroke: ",s
250 INPUT AT(10)"Rod length (if not known, enter 0): ",l: IF l=0 THEN l=2*s
260 a=b^2*PI/4
270 d=a*s: REM area and displacement for ONE piston (in mm2, mm3)
280 PRINT: PRINT AT(10)"To alter, press A: to continue press any other key"
290 REPEAT: a$=INKEY$: UNTIL a$>"":IF a$="a" OR a$="A" THEN 205
295 GOSUB spec
300 RETURN


400 LABEL inpb
410 CLS
420 p=0: pb=0
430 INPUT AT(10;2) "Port height: ",h
440 INPUT AT(10) "Port width:  ",w
450 PRINT: PRINT: PRINT AT(10) "If the following data is not known, enter 0": PRINT
460 INPUT AT(10)"Intake opens, mm BTDC: ",po
470 IF po=0 THEN INPUT AT(10)"Intake opens, degrees BTDC: ",p: IF p=0 THEN PRINT
AT(10)"Distance from bottom edge of port": INPUT AT(10)"to top edge of piston at TDC, in
mm: ",pb
480 IF p=0 AND po=0 AND pb=0 THEN PRINT: PRINT AT(10)"Insufficient data.":PRINT
AT(10)"Enter port timing or piston/height dimension.": GOTO 460
490 PRINT: PRINT
500 PRINT AT(10)"Piston dimensions, in mm:": PRINT
510 INPUT AT(10)"Piston pin bore: ",z
520 INPUT AT(10)"Distance from top of pin to edge of crown: ",x
540 PRINT AT(10)"Distance from edge of crown to the bottom of the skirt,"
560 INPUT AT(10)"in the position of the port: ",y
570 f=x+z/2: g=y-f: REM g - distance of skirt below piston pin centre
580 GOSUB calc
590 PRINT: PRINT: PRINT TAB(31)"Press A to alter data, or C to continue."
600 REPEAT: a$=INKEY$: UNTIL a$>"": IF a$="a" OR a$="A" THEN 410
605 GOSUB spec
610 RETURN


700 LABEL menu
710 CLS
720 PRINT AT(16;5)"Time-area against.........crank speed (1)"
730 PRINT AT(35)"........port width (2)"
740 PRINT AT(35)".....port top edge (3)"
750 PRINT AT(35)"..port bottom edge (4)"
760 PRINT AT(35)"piston skirt length(5)"
770 PRINT AT(16)"Change engine dimensions..............(6)"
780 PRINT AT(16)"Change piston/port dimensions.........(7)"
790 PRINT AT(16)"STOP................................(8)"
800 PRINT: PRINT AT(16)"To see current engine specification, use Window #2"
810 PRINT: PRINT AT(16)"Option: ": REPEAT:mm=INKEY: UNTIL mm>-1: m=mm-48
820 ON m GOSUB sp,w,t,b,sk,edim,pdim,830
830 RETURN


900 LABEL calc
910 IF po>0 THEN 940
920 IF p>0 THEN c=s/2*COS(p)+SQR(l^2-(s/2*SIN(p))^2): po=s/2+l-c
930 IF p=0 THEN po=pb-y
935 po=ROUND(po,2)
940 PRINT:PRINT:PRINT TAB(10)"Piston length: "y
```

160

```
950 PRINT TAB(10)"Intake opens: "po"mm BTDC."
960 IF pb>0 THEN 980
970 pb=po+y: IF p>0 THEN 1020
980 IF p=0 THEN PRINT "Please wait";: REPEAT: q=s/2*COS(p)+SQR(l^2-
(s/2*SIN(p))^2):p=p+1:  UNTIL  q<s/2+l-po:  p=p-2
990  REPEAT:p=p+0.005:  q=s/2*COS(p)+SQR(l^2-(s/2*SIN(p))^2)
1000  UNTIL  q<s/2+l-po

1010 p=ROUND(p,2): PRINT TAB(24)p" deg BTDC"
1020 PRINT: PRINT TAB(30)"Press any key to continue."
1030 REPEAT: a$=INKEY$: UNTIL a$>"": GOSUB spec
1040 RETURN


1100 LABEL edim
1110 GOSUB deck
1120 CLS
1130 GOSUB inpa
1140 GOSUB calc
1150 RETURN


1200 LABEL pdim
1210 GOSUB deck
1220 CLS
1230 GOSUB inpb
1240 RETURN


1300 LABEL deck
1310 CLS
1320 PRINT AT(10;4)"If the stroke, rod length or piston dimensions"
1330 PRINT: PRINT AT(10)"are changed, then the program assumes"
1340 PRINT: PRINT AT(10)"that the barrel will be moved to match"
1350 PRINT: PRINT AT(10)"the new deck height. The datum for all"
1355 PRINT: PRINT AT(10)"measurements is the piston height at TDC."
1360 PRINT AT(30;16)"Press any key to continue."
1370 REPEAT: a$=INKEY$: UNTIL a$>""
1380 RETURN


1400 LABEL ta
1410 i=1 : REM adjustable
1420 t=i/n:  ta=0
1430 FOR e=0 TO 360 STEP i: REM degrees
1440  c=s/2*COS(e)+SQR(l^2-(s/2*SIN(e))^2)
1450  c1=s/2+l-c+y
1460 h1=pb-c1: IF h1>h THEN h1=h
1470 IF h1<0 THEN h1=0
1480  ta=ta+t*h1*w:
1490 NEXT
1500 RETURN


1600 LABEL sp
1610 CLS
1620 INPUT AT(7)"Minimum crank speed, rev/min: ",n1: n1=n1*6: REM deg/s
1630 INPUT AT(7)"Maximum crank speed, rev/min: ",n2: n2=n2*6
1640 INPUT AT(7)"Increment: ",n3: n3=n3*6
1650 PRINT: PRINT "Crank","Time-area","Specific t-a"
1660 PRINT "rev/min","s-sq mm","s-sq mm/cc x10^-3": PRINT
1670 FOR n=n1 TO n2 STEP n3
```

161

```
1680 GOSUB ta
1690  PRINT n/6,ROUND(ta,3),ROUND(ta/d*10^6,3)
1700 GOSUB wait
1710 NEXT
1720 PRINT: PRINT TAB(15)"Press any key to continue."
1730 REPEAT: a$=INKEY$ UNTIL a$>""
1740 RETURN


1800 LABEL wait
1810 IF aa>0 THEN aa=aa+1: GOTO 1870
1820 yy=YPOS
1830 IF yy>750 THEN 1880
1840 PRINT AT(10;20)"Press SPACE bar to continue.";
1850 IF INKEY$<>" " THEN 1850
1860 aa=1
1870 IF aa=13 THEN aa=0: PRINT: PRINT: GOTO 1840
1880 RETURN


1900 LABEL w
1910 j=w
1920 CLS
1930 PRINT AT(10)"Current port width is "w"mm.
1940 INPUT AT(10)"Minimum width: ",w1
1950 INPUT AT(10)"Maximum width: ",w2
1960 INPUT AT(10)"Increment:      ",w3
1970 INPUT AT(10)"Crank speed, rev/min: ",n:n=n*6
1980 PRINT: PRINT "Width, mm","Time-area","Specific t-a"
1990 PRINT " ","s-sq mm","s-sq mm/cc *10^-3"
2000 PRINT
2010 FOR w=w1 TO w2 STEP w3
2020 GOSUB ta
2030  PRINT w,ROUND(ta,3),ROUND(ta/d*1E6,3)
2040 GOSUB wait
2050 NEXT
2060 w=j: PRINT
2065 GOSUB wait

2070 PRINT TAB(8) "Current port width is "w"mm."
2080 PRINT TAB(8)"Enter required value: ";:INPUT"",w
2090 PRINT: PRINT TAB(30)"Press any key to continue."
2100 REPEAT: a$=INKEY$ UNTIL a$>"": GOSUB spec
2110 RETURN


2200 LABEL t
2210 CLS
2220 j=h
2230 PRINT AT(8)"Port height is "h"mm."
2240 PRINT AT(8)"Vary time-area by moving the top edge of the port."
2250 INPUT AT(8)"Minimum port height: ",h1
2260 INPUT AT(8)"Maximum port height: ",h2
2270 INPUT AT(8)"Increment: ",h3
2280 INPUT AT(8)"Crank speed, rev/min: ",n: n=n*6
2290 PRINT
2300  PRINT "Height, mm","Time-area","Specific t-a"
2310  PRINT " ","s-sq mm","s-sq mm/cc x10^-3"
2320 PRINT
2330 FOR h=h1 TO h2 STEP h3
```

162

```
2340 GOSUB ta
2350 PRINT h,ROUND(ta,3),ROUND(ta/d*1E6,3)
2360 GOSUB wait
2370 NEXT
2380 h=j
2390 PRINT: PRINT TAB(8)"Current port height is "h"mm."
2400 PRINT TAB(8)"Enter the required height: ";:INPUT " ",h
2410 PRINT: PRINT TAB(30)"Press any key to continue."
2420 REPEAT: a$=INKEY$: UNTIL a$>"":GOSUB spec
2430 RETURN


2500 LABEL b
2510 CLS
2520 j=pb: k=h
2530 PRINT AT(8)"Port height is "ROUND(h,2)"mm."
2535 PRINT AT(8)"The bottom edge is "ROUND(pb,2)"mm below TDC."
2540 PRINT AT(8)"Vary time-area by moving the bottom edge of the port."
2560 INPUT AT(8)"Highest port position from TDC: ",pb1
2570 INPUT AT(8)"Lowest port position from TDC: ",pb2
2580 INPUT AT(8)"Increment: ",pb3
2590 INPUT AT(8)"Crank speed, rev/min: ",n : n=n*6
2600 PRINT: PRINT "Posn BTDC","Height","Time-area","Specific t-a"
2610 PRINT "mm","mm","s-sq mm","s-sq mm/cc x10^-3"
2620 PRINT
2630 FOR pb=pb1 TO pb2 STEP pb3
2640 GOSUB wait
2650 h=k+pb-j
2660 GOSUB ta
2670 PRINT ROUND(pb,2),ROUND(h,2),ROUND(ta,3),ROUND(ta/d*1E6,3)
2680 NEXT
2690 h=k: pb=j
2700 PRINT: PRINT TAB(8)"Current position of bottom edge is "ROUND(pb,2)"mm BTDC."
2710 PRINT TAB(8)"Enter required position: ";: INPUT" ",pb: h=h+pb-j
2720 po=0:p=0
2730 GOSUB calc
2760 RETURN


2800 LABEL spec
2810 WINDOW #2 SIZE 35,17
2820 WINDOW #2 TITLE "Current specification"
2830 WINDOW #2 PLACE 376,-24
2835 CLS #2
2840 PRINT #2 "Bore x stroke, mm        "b"x"s
2850 PRINT #2 "One piston area, mm^2  "ROUND(a,2)
2860 PRINT #2 "One piston dispt, cm^3  "ROUND(d/1000,2)
2870 PRINT #2 "Rod length, mm           "l
2880 PRINT #2 "Piston pin dia, mm        "z
2890 PRINT #2 "Piston skirt length, mm    "y
2900 PRINT #2 "Deck ht from top of pin, mm "x
2910 PRINT #2 "Intake opens, mm BTDC      "ROUND(po,2)
2920 PRINT #2 "Intake opens, deg BTDC      "ROUND(p,2)
2930 PRINT #2 "Intake duration, deg      "ROUND(2*p,2)
2940 PRINT #2 "Port height, mm          "ROUND(h,2)
2950 PRINT #2 "Port bottom edge, mm BTDC  "ROUND(pb,2)
2960 PRINT #2 "Port width, mm           "w
2970 PRINT
2980 RETURN
```

```
3000 LABEL sk
3010 CLS
3020 j=y
3030 PRINT AT(8)"Piston skirt length is "ROUND(y,2)"mm."
3040 PRINT AT(8)"Alter time-area by changing the skirt length."
3050 INPUT AT(8)"Maximum skirt length: ",y1
3060 INPUT AT(8)"Minimum skirt length: ",y2
3070 INPUT AT(8)"Increment: ",y3
3080 INPUT AT(8)"Crank speed, rev/min: ",n: n=n*6
3090 PRINT
3100 PRINT "Length, mm","Time-area","Specific t-a"
3110 PRINT " ","s-sq mm","s-sq mm/cc x10^-3"
3120 PRINT
3130 FOR y=y1 TO y2 STEP -y3
3140 GOSUB wait
3150 GOSUB ta
3160 PRINT y, ROUND(ta,3),ROUND(ta/d*1E6,3)
3170 NEXT
3180  y=j
3190 PRINT
3200 PRINT TAB(8)"Current skirt length is "y"mm."
3210 PRINT TAB(9)"Enter required length: ";: INPUT "",y
3220 p=0: po=0
3230 GOSUB calc
3240 RETURN
```

In disc valve engines the time-area is important, and is fairly easy to calculate, as shown below. However the timing of the port closing is equally important, as this has a major effect on the power characteristics. Basically, later closing corresponds to power at higher revs.

To calculate the disc valve time-area, assume that the port opens when the opening edge of the disc bisects the port area and that it closes when the closing edge bisects the port area. This takes a simple average of the port area during the opening and closing phases; the rest of the time, the port is fully open.

$$\text{time-area (disc valve)} = AX/b$$

$$\text{Where} \quad A = \frac{Y}{2}(R2 - R1)(R2 + R1) \quad \text{(wedge shaped)}$$

$$\text{or} \quad A = pq \quad \text{(rectangular shaped port)}$$

$$\text{or} \quad A = 3.14\,r^2 \quad \text{(circular shaped port)}$$

Where:
X — open angle of disc in radians
b — crank speed in radian/second
Y — angular opening of port (radians)
R1 — radius of port inner edge, mm
R2 — radius of port outer edge, mm
p, q — sides of rectangular port, mm
r — radius of circular port, mm

Note: to convert degrees into radians, multiply by 0.01745.

164

The time-area program offers the facility to compare an engine with the latest available machinery, so it is possible for it to stay totally up to date. However it seems that some figures are fairly timeless; people have been using the time-areas shown below for at least 15 years, and they still work:*

intake   0.015 to 0.020 s-mm^2/cm^3
scavenge   0.008 to 0.010 s-mm^2/cm^3
exhaust   0.014 to 0.016 s-mm^2/cm^3

* Add up to 25% for a GP racer.

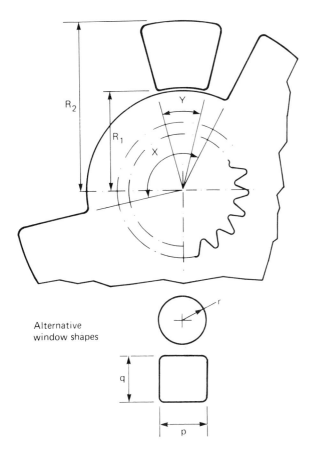

Fig. 86. Dimensions used in the time-area calculation for disc valves and different port shapes

Alternative
window shapes

Piston motion

This program uses computer graphics to animate the motion of a crank and piston, demonstrating the simple phasing of intake, scavenge and exhaust processes. Although the engine is crudely simplified, a visual guide is helpful if following the overlapping cycles.

165

```
  10 REM PISTON
  20 REM 8/6/85
  30CLS
  40 MODE 0
  45 VDU 23,1,0;0;0;0;
  50 MOVE 600,400
  60 FOR C=0 TO 6.3 STEP 0.01
  70 DRAW 101*SINC+600,101*COSC+200
  80 NEXT
  90 MOVE 805,850
 100 DRAW 805,1000
 110 DRAW 395,1000
 120 DRAW 395,805
 130 DRAW 300,805
 140 DRAW 300,400
 150 MOVE 395,750
 160 DRAW 395,400
 170 MOVE 805,850
 180 DRAW 1000,850
 190 MOVE 1000,700
 200 DRAW 805,750
 210DRAW 805,700
 220 DRAW 1000,680
 230 MOVE 1000,600
 240 DRAW 805,620
 250 DRAW 805,0
 260 DRAW 300,0
 270 DRAW 300,400
 280    FOR A=0 TO 100000 STEP 0.1
 290 IF C<800 THEN VDU 31,65,7:PRINT"EXHAUST"
 300 IF C>800 THEN VDU 31,65,7:PRINT"       "
 310 IF C>770 THEN VDU 31,65,11:PRINT"INTAKE"
 320 IF C<770 THEN VDU 31,65,11:PRINT"            "
 330 IF C<750 THEN VDU 31,7,10:PRINT"SCAVENGE"
 340 IF C>750 THEN VDU 31,7,10:PRINT"         "
 350 MOVE 600,200
 360 DRAW 500,200
 370 DRAW 700,200
 380 MOVE 600,200
 390 DRAW 600,100
 400 DRAW 600,300
 410X=SINA
 420Y=COSA
 430 MOVE X*100+600, Y*100+200
 440 C=SQR(ABS(800^2-10000*X^2))+100*Y
 450 DRAW 600, C-50
 460 MOVE 800, C+50
 470 DRAW 400,C+50
 480 DRAW 400,C-150
 490 DRAW 800,C-150
 500 DRAW 800,C+50
 510 B=A-0.1
 520 P=SINB
 530 Q=COSB
 540 MOVE P*100+600, Q*100+200
 550 D=SQR(ABS(800^2-10000*P^2))+100*Q
 560 PLOT 7,600,D-50
 570 PLOT 7,800,D+50
 580 PLOT 7,400,D+50
 590 PLOT 7,400,D-150
 600 PLOT 7,800,D-150
 605 PLOT 7,800,D+50
 610    MOVE 100*SIN(B-0.1)+600,100*COS(B-0.1)+200
 620 DRAW P*100+600, Q*100+200
 630 NEXT
 640 GOTO 170
 650 END
```

166

Index